8/98

THE HIDDEN SEA

THE HIDDEN SEA

Ground Water, Springs, and Wells

Francis H. Chapelle

Illustrated by

James E. Landmeyer

Library of Congress Catalog Card Number 97-5640
ISBN 0-945005-26-1

Published by Geoscience Press, Inc.
P.O. Box 42948
Tucson, AZ 85733-2948

Manufactured in the United States of America
Printed on acid-free paper.
10 9 8 7 6 5 4 3 2 1

Cover design by Margaret Donharl, Boulder, Colorado
Book design by Dianne Nelson, Golden, Colorado

LIBRARY OF CONGRESS CATALOGING-IN-PUBLICATION DATA

Chapelle, Frank.
The hidden sea : ground water, science, and environmental realism / Frank H. Chapelle ; illustrations by James E. Landmeyer.
p. cm.
Includes bibliographical references and index.
ISBN 0-945005-26-1
1. Groundwater. I. Landmeyer, James E. II. Title.
GB1003.2.C425 1997 97-5640
553.79 — dc21 CIP

For Adrienne,

lithe and lovely

CONTENTS

PREFACE

WHEN ABRAHAM REACHED THE PROMISED LAND some 3,900 years ago, he found that most of the arable land in the Jordan Valley or on the coastal plain was already firmly occupied by the Canaanites. Furthermore, these strong and warlike people had no intention of sharing this land with some upstart nomad just out of the desert. If Abraham wanted to make Canaan his home, he would have to colonize the desolate Judean hill country, which nobody else particularly wanted. Abraham was a shepherd by trade, and these rocky hills presented problems for anybody trying to raise sheep. The biggest problem was finding a reliable water supply. Sheep need water in order to thrive—and streams, rivers, and ponds were singularly lacking in these arid hills.

But to Abraham, who had learned the art of digging wells while he was wandering through Mesopotamia on his way to Canaan, the lack of streams or ponds in the hill country was just an inconvenience. Within a few years, he had dug a series of wells and instituted a system of moving his sheep from well to well so that grass was always abundant. Under this skillful husbandry, Abraham's flocks multiplied rapidly. Eventually, and at least partly because of his skill in locating and digging wells, Abraham became a wealthy and powerful man.

GROUND WATER, that vast ocean of water lying beneath the earth's surface, has long been a source of wealth to humankind. Springs and wells have supplied fresh drinking and irrigation water throughout history, and this is as true today as it was in

Abraham's time. In the United States, one-third of the population—100 million people—depends on ground water for its everyday needs. Also, much of the agricultural production in the United States would be impossible without irrigation wells and the ground water they produce.

But ground water is more than just a useful commodity. Because it lies hidden in the secret depths of the earth, ground water has always fired the human imagination. Where does it come from? How does it move? Where do you find it? Ancient people developed elaborate mythologies—subterranean spirits, wishing wells, water witches—to answer these questions, and these mystical traditions remain deeply embedded in our culture. More recently, people have developed intricate geological and mathematical models to address these very same questions. While there is little outward similarity between these mystical and rational approaches to understanding ground water, they are both the offspring of human imagination.

This is a book about how both mysticism and rationality have been used to understand the puzzling behavior of ground water (part 1). It is a book about the strikingly different characteristics of ground water systems found in different parts of the country, and how these differences affect the people who use them (part 2). It is about how ground water systems can become contaminated by various waste disposal practices, and how perceptions of this contamination are so often different from reality (part 3). But even more, this is a book about how human beings go about discovering the hidden secrets of the earth. It is about human imagination.

—Francis H. Chapelle

Part One

Myths and Models

In the mind of early man, water was not acted on by laws of force, but by life and will; the water-spirits of primeval mythology are as souls which cause the water's rush and rest, its kindness and its cruelty.

E.B. Taylor, Nineteenth Century Antiquinarian

Chapter One

The Hidden Sea

"My well is poisoned," the elderly woman said simply, "it's been poisoned by that dump." Although her voice was even, the woman's eyes were hard with anger. The dump she was referring to was a state-of-the-art secure landfill located about a mile from her house. Landfills, "secure" or not, are notorious for leaking noxious chemicals and contaminating nearby wells. This woman was convinced that this had happened to her.

It was just chance that Tracey Kean, a hydrologic technician with the U.S. Geological Survey (USGS), had stopped at this particular woman's house in the first place. The USGS was conducting a study on the availability and chemical quality of ground water in the vicinity of Baltimore, Maryland. An important part of this study was to prepare an inventory of selected wells in the area, locate them accurately on maps, and obtain water samples for chemical analysis. In Anne Arundel County, which is just south of Baltimore, many homes have individual wells that are used for domestic water supply. Tracey's job was to locate at least one well per square mile, add it to the inventory, and, with the permission of the owners, take a water sample.

This was something that Tracey was particularly good at. Most homeowners become understandably suspicious when some stranger drives up in a truck and starts asking questions about their wells. But Tracey was such a friendly and cheerful person she generally managed to extract cooperation from even the grumpiest homeowners. There were several wells in this general area that Tracey could have picked. But, for whatever reason, she had chosen this one.

"Our water has gotten worse in the last few years," the woman continued angrily. "Now the only thing we use it for is flushing the toilet. We have to buy bottled water for drinking and cooking. It has to be the dump."

On the face of it, this was unlikely in the extreme. The landfill in question was fairly new and belonged to Browning Ferris Industries (BFI), one of the largest solid waste management companies in America. As such, it conformed to all sorts of rules and regulations designed to prevent any sort of leak. Plus, it was a full mile away from the woman's well. A mile is an awfully long way for contaminated ground water to migrate from a landfill.

But it wasn't impossible, and if the woman's well did happen to be contaminated, it could indicate a much larger contamination problem. The only thing Tracey could do was to presume the woman was right, and check out the well as thoroughly as possible. Regardless of what happened, at least Tracey would get the information she needed for the well inventory.

"Well, if the landfill is really leaking, that would be pretty serious," she said with a frown of concern, "I could tell you more if I were able to examine your well. Would that be possible?" The woman hesitated for a second and looked at Tracey doubtfully. Finally she motioned for her to come around the side of the house.

The woman's well was made out of standard 4-inch PVC casing and was located behind her house. It took Tracey a few minutes to remove the cap, which was held firmly in place by three bolts. But once the cap was off, it was quickly apparent that the woman's complaints were not imaginary. Wafting from the well was the strong odor of hydrogen sulfide gas. Hydrogen sulfide gives off an unpleasant "rotten egg" smell and can be an indication of polluted water. It was easy to see why this woman was so sure that her well had been contaminated. Furthermore, given the proximity of the landfill, it was easy to see why she would think it was the source of her troubles.

"Phew," Tracey said with a grimace as she turned her head away, "there's no arguing about the smell."

"Like I said," the woman said grimly "the water's poisoned."

"It could be," Tracey replied. For the first time, she was becoming concerned. There was no doubt that something was causing the foul smell, and if this well was really contaminated by landfill leachate, it could be a very, very serious situation indeed.

Tracey turned to face the woman, "If you'd like me to take a sample of the water, I'll have it analyzed so we can figure out what might be wrong." When the woman hesitated, Tracey quickly added, "It won't cost you anything, and I'll make sure you get a copy of the analysis." After another second of hesitation, the woman nodded in agreement.

Tracey replaced the well cap and returned to her truck for some sample bottles. Next, she turned on the outside tap and let the water run until she heard the submersible pump in the well turn on with a gentle click. After the pump had hummed along for a few minutes, she filled the sample bottles, adding the prescribed preservatives, and capped them. Tracey thanked the woman for her cooperation, promised to send her a copy of the analysis, and drove away in her truck.

THERE ARE ABOUT 100 MILLION AMERICANS who, like this elderly lady from Maryland, depend upon ground water for day-to-day use.

Many of these people, particularly those who own their own wells, are acutely aware that ground water can become contaminated from a variety of sources. Landfills, septic fields, waste disposal pits, agricultural chemicals, and leaking underground fuel tanks are just a few of the contrivances that regularly contaminate wells.

In some cases when a well becomes contaminated, it is only too obvious by the smell or taste of the water. Landfill leachate, for example, is notoriously foul-smelling. Some contaminants, however, such as pesticide residues or other organic chemicals, may give no outward sign at all of their presence. It is this sort of hidden contamination that is the most dangerous. There are many examples of this potentially deadly contamination, but perhaps the most tragic was the case of Woburn, Massachusetts.

Woburn is situated in the valley of the Aberjona River. This valley, like many in the northeastern United States, was formed when the great ice sheets covering the North American continent began to melt about 15,000 years ago. Water from the melting glaciers collected into large rivers and eroded steep valleys into the granite-gneiss bedrock. Once the glaciers were gone, however, the incised valleys gradually filled with layers of sand, gravel, and silt. The sediments that fill these "buried valleys" are typically about 100 feet thick, and are very permeable. They are some of the most productive sources of ground water in the northeastern United States.

This readily available ground water is a boon to towns like Woburn because it is a cheap, convenient source for municipal water supply. In the 1960s and 1970s, Woburn was able to provide all of its water needs from ten or so municipal wells spaced throughout the town. Whenever more water was needed, all the townspeople had to do was drill another well.

But these permeable valley-fill sediments had other useful properties as well. Woburn is an old industrial town, and, over the years, various manufacturing plants made products such as chemicals, machine tools, and paints. Manufacturing these products generated liquid wastes such as paint thinners (trichloroethene or

TCE) and metal degreasers (perchloroethene or PCE). Getting rid of these wastes was a constant nuisance. However, it was soon discovered that because the soils and subsurface sediments were so permeable, wastes dumped into shallow holes quickly disappeared. To the average manager of an industrial plant in the 1960s, this was the ideal solution to the vexing waste disposal problem.

For years, the buried-valley aquifer underlying Woburn served as a source of drinking water *and* as a repository for industrial wastes. Most of the people in the town were unaware that any of this was going on, and for many years there were no apparent problems with these dual practices.

The first indication that something might be wrong came in the late 1970s. Some local citizens noticed that an unusual number of children in East Woburn were coming down with childhood leukemia, and several of these children eventually died. Painstaking investigation finally established that the incidence of leukemia was associated with drinking water from two municipal wells. Analyses of the well water showed that the water was contaminated by TCE and PCE.

Until the contamination problem was discovered, the hidden nature of the ground water system underlying Woburn had seemed either innocuous (to the municipal well operators) or actually advantageous (to the industries disposing liquid wastes). Now, however, it was evident that this hidden nature could have a sinister side as well. The seeping organic chemicals, which were drawn inexorably toward the pumping municipal wells like rainwater to a drain, were effectively concealed from direct observation. Since these contaminants lacked any taste or smell, their eventual arrival at the wells went unnoticed for years,

The fact that it took so long to discover the contamination made the situation considerably worse. While people are still arguing whether the TCE caused leukemia to develop in the unfortunate children, nobody disputes the notion that chronic exposure to TCE in drinking water *might* have toxic effects. Thus the fact that the contamination remained hidden and undetected for so long made a potentially bad situation considerably worse. This tragic incident was caused not only by human incompetence in disposing of the toxic wastes. A significant component was the hidden, secret nature of the underlying ground water system itself.

HORROR STORIES LIKE WOBURN, MASSACHUSETTS, are the ones that most often get in the newspapers. However, the concealed nature of ground water systems can lead to other problems as well.

Take, for example, the elderly woman from Maryland whose well Tracey Kean had sampled. After sampling the well, Tracey returned to the office, did the necessary (and dreary) paperwork, and shipped the samples off to the lab. Because leakage from the landfill was suspected, she asked the lab to scan for a variety of dissolved organic carbon compounds including TCE. In addition, she asked them to look for dissolved nitrogen compounds such as nitrate and ammonia, commonly found in landfill leachate. If contaminated water were really reaching the woman's well from the landfill, it should show up in these tests.

A month later, Tracey got the lab results. What the analysis showed, however, was that the water was absolutely pristine. There were no organic compounds normally associated with landfill leachate, there were no nitrogen compounds, and in fact there was very little in the way of dissolved solids at all. The elderly woman's well water was as clean as the proverbial whistle.

So where did the awful rotten egg smells in her well come from?

The answer to that question hinged on something that had happened 120 million years ago. This woman's well tapped sediments of the Patapsco formation, which was laid down by an ancient river during Cretaceous time. This ancient river, like modern rivers, was prone to picking up tree trunks, carrying them along for a way, and finally burying them under layers of silt, sand, and clay. It is not at all uncommon to encounter these buried logs while drilling wells into the sands of the Patapsco formation.

If you want a well for drinking water, however, encountering these old tree trunks during drilling is bad luck. Even though the wood in these tree trunks is more than 100 million years old, the organic matter they are made out of is still being slowly degraded by bacteria. As this degradation proceeds, it transforms dissolved sulfate (SO_4) present in the ground water into hydrogen sulfide (H_2S). Hydrogen sulfide is harmless in trace amounts, but it can make water smell really bad. The problem with this woman's well had nothing to do with the BFI landfill, and nothing to do with ground water pollution. She was just unlucky enough to have drilled her well too near the final resting place of an ancient tree trunk (figure 1.1).

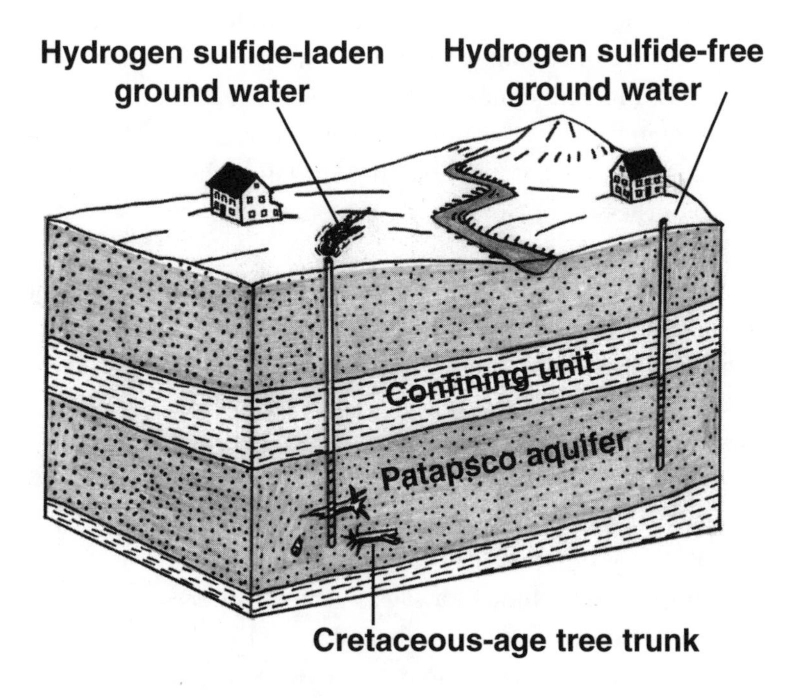

Figure 1.1. How the presence of buried tree trunks in the Patapsco formation controls the distribution of hydrogen sulfide-laden ground water.

When the lab results came in, Tracey called the elderly woman and told her what the problem was. She assured the woman that her well water, while unpleasantly fragrant, was perfectly safe to drink. Furthermore, with the proper water treatment, it would be more than satisfactory for drinking, cooking, or any other purpose.

The elderly woman, however, was unimpressed. "That dump has poisoned my well," she insisted, and there was nothing Tracey could say to change her mind. The BFI landfill, after all, was right out there in plain sight. The ancient tree trunks that were the real source of her troubles were buried far out of sight and mind.

In the case of Woburn, Massachusetts, the hidden nature of the underlying ground water system was responsible for letting a potentially life-threatening contamination event go unnoticed for years. For the woman in Maryland, the situation was exactly reversed. The hidden nature of the ground water system resulted in the *appearance* of contamination when in fact none was present. In both cases, this hiddenness extracted costs. In Woburn, the cost was the contamination of a municipal water supply, and possibly the illnesses of innocent people. In Maryland, the cost was the needless bitterness of one elderly woman.

GROUND WATER, that vast body of water underlying the visible world, is a great Hidden Sea. Many people believe that most of the fresh water in the world is stored in surface water bodies—lakes, rivers, and streams—that are readily observable and subject to arbitrary human control. This just isn't so. Of the one or so quintillion (which is to say, one billion billion) gallons of fresh liquid water present in North America, a mere 2 percent resides in rivers or lakes at any given time. By far the largest portion—fully 980 quadrillion gallons—is hidden away as ground water.

This vast amount of water has been both a blessing and a curse for humankind ever since people first began to scratch shallow wells with their bare hands. By finding and tapping hidden sources of ground water, it has been possible to transform unproductive wastelands into thriving farms and cities. But on the other hand, humans have often suffered misery and ruin when the wells dried up or became contaminated.

But even more, ground water has always been a source of enduring mystery. Where did it come from? How does it move? Where can you find it? To ancient people, these were literally matters of life and death. But because ground water was hidden from their view, the only answers they found came from their own imaginations. Wells were believed to be inhabited by spirits, some of whom would grant secret wishes if offered gifts. Some springs were thought to produce holy water that could heal a variety of human ailments. Others that produced salty or sulfurous waters were thought to be under the sway of the devil. These mystical traditions—which dominated human thought for centuries—are images formed by the lens of human imagination.

The story of the Hidden Sea properly begins with these mystical traditions. This is not because these traditions are quaint and amusing. It is because they have a history that is thousands of years old, and because they still have a powerful impact on how people think about ground water, springs, and wells.

Consider, for example, the coins in the fountains.

Additional Reading

WGBH (Boston). 1986. Transcript of *Nova* episode 1306. "Toxic Trials."

Landmeyer

CHAPTER TWO

Wells of Hope

THE FOUNTAIN COULD BE JUST ABOUT ANYWHERE IN AMERICA—a mall or an airport or a hotel lobby. It might be a lavish affair, with water bubbling out of multiple ports and coursing swiftly over sluices and waterfalls. Or it might be a single spigot with a plain, unadorned basin. On the one hand, the popularity of fountains simply reflects the pleasure people get from watching and listening to running water. On the other, it also shows that some ancient customs surrounding springs and wells are still very much alive. The evidence for this is plain enough. It is the small change—pennies, nickels, and dimes—that inevitably accumulates in the fountain basins.

The custom of leaving small gifts and offerings at wells and springs, with the hope of having "good luck" in return, is a very old and very deeply ingrained custom in many cultures. The origins of these customs are misty, and certainly predate Christianity.

But it is interesting to compare the kinds of offerings now in vogue (pennies in a fountain) to those that used to be made at springs and wells.

In nineteenth-century Scotland, for example, a well in Peeblesshire was popularly referred to as the "Cheese Well." The local custom, which had been carried on for centuries, was for passersby to drop a piece of cheese into the well as offerings to the water spirits so that they would provide good luck. In Northumberland, a well went by the name of the "Rag Well" because the spirits seemed to like bits of rag tied to the bushes nearby. Deities inhabiting other wells were reputed to desire small coins, colored pebbles, bits of metal, and even pins. One particular well in Northumberland was believed to be inhabited by a sprite whose special calling was easing the daily chores of women. It became the custom for young girls to drop in a pin when passing the well in order to gain the favor of the fairy—favor they might need when burdened by the heavy labor of married life.

The practice of making offerings at wells was not confined to the British Isles by any means. The religion of the ancient Romans, for example, was based on the belief that natural objects were inhabited by spirits. These spirits could choose to help or hinder human endeavors at their whim. To secure their help, the Romans gave offerings (*sacrae stipes*) at places where the spirits dwelt. Lakes, rivers, springs, and wells were favorite haunts of the spirits, and it was there that the Romans came with their offerings.

The archaeological evidence for these practices comes from a variety of sources. For example, in 1852 some Jesuit fathers in Italy, who were the caretakers of a sulfur spring known as Sorgenti de Vicarello, resolved to clean and repair the ancient masonry that surrounded the well. They brought in a team of masons from Rome to drain the well by diverting the upwelling water through a conduit, and make the necessary repairs. At the bottom of the well, the workmen came across a layer of brass and silver coins dating to the fourth century after Christ. Lower down, they came across a layer of gold and silver coins from the reign of Augustus. Successive layers of coins were discovered in deeper horizons that dated to before the founding of Rome. The deepest stratum had a layer of arrowheads and knives made of polished stone. Quite clearly, this spring had been the site of offerings for thousands of years.

Of course we don't really know what the ancient Romans were hoping for in return for their offerings. But the custom of the wishing well, where a wisher would provide an offering in the hope of receiving a specific favor, is very common in folk traditions. St. Anthony's Well in Edinburgh, Scotland, is one of the most famous, and the ritual there seems to have been widely copied. This was to drop an offering into the well while quietly reciting your wish. If, however, you told anyone of your wish, it would never come true. It is a pretty good bet that many of the coins found in shopping mall fountains represent a child's wish for some new toy.

While the St. Anthony's custom is probably the most commonly remembered, some wishing wells had more elaborate rituals associated with them. In Aberdeenshire, Scotland, for example, it once was the custom for a bride, on the eve of her marriage, to go to the Bride's Well. There her bridesmaids would bathe her feet with water drawn from the well to ensure that she would be able to bear children. The bride then tossed a few crumbs of cheese and bread into the well so that her children, once born, would never go hungry.

Many wishing wells dealt specifically with matters of the heart. Some wells, for example, were reputed to be able to show the face of your future spouse. The technique involved dropping a coin into the well, and waiting for a vision of the person's face to appear in the ripples. This, presumably, would help the coin dropper to find the spouse who had been ordained by fate.

In the case of the British Isles, the traditions of wishing wells and making offerings at wells originated outside of Christianity. This fact was widely noted by the nineteenth-century antiquinarians who recorded the folklore surrounding wells and springs. E. B. Tylor, for example, wrote:

> What ethnography has to teach us of that great element of the religion of mankind—the worship of well and brook, is simply this—in the mind of early man, water was not acted on by laws of force, but by life and will; that the water-spirits of primeval mythology are as souls which cause the water's rush and rest, its kindness and its cruelty; that, lastly, man finds . . . deities with a wider influence over his life, deities to be feared and loved, to be prayed to and praised, and propitiated with sacrificial gifts.

Reverence for water and springs is very common and can be found at some level in most religions. It is reasonable to wonder, therefore, just how these traditions get started. In most cases, trying to trace a tradition to its beginnings is simply impossible—such events were seldom recorded. In a few tantalizing cases, however, it is possible to make sense of an ancient tradition, and these cases shed considerable light on how water reverence begins.

One of the few recorded examples of how springs become associated with a deity comes from the Bible, where the occurrence of springs and wells is often interpreted as reflecting the intervention of God. During the Exodus, when the Israelites had finally escaped from Egypt and were wandering through Sinai, they came upon a particularly bleak desert known as the Wilderness of Sin. While passing through this barren place, the people became desperate for lack of water, and Moses feared an open rebellion. Exodus 17:11 records:

> And the people thirsted there for water; and the people murmured against Moses, and said, Wherefore is this that thou hast brought us up out of Egypt, to kill us and our children and our cattle with thirst?

According to the Bible, the Lord tells Moses to take his rod—the same one he used to part the Red Sea—and strike a rock.

Moses obeys, and a stream of water promptly issues from the rock. The people were able to drink their fill and, at least temporarily, ceased their grumbling. The Bible interprets this event as a miracle, where God intervenes personally to save His people.

This, however, was no miracle.

Much of the Sinai Peninsula is underlain by limestone. In places, ground water flowing through cracks and fissures in the limestone reach land surface, forming a spring. Unlike the famous oases found in the Arabian Desert, which produce large quantities of water from a deeply buried sandstone, most of these limestone springs are little more than seeps with very little water flowing from them. In the dry heat of the desert, the water from these low-discharge springs evaporates very rapidly, and minerals dissolved in the water are left behind to form a crust on the limestone. In some cases this crust seals up the cracks, and the water accumulating behind this "seal" often builds up considerable pressure.

Modern desert dwellers, principally Bedouins, know how to recognize these plugged springs for what they are and how to release the water by breaking the mineralized seal with a sharp blow. Ancient desert dwellers, such as the Midianites among whom Moses had lived for years before he led the Israelites out of Egypt, apparently knew this as well. When Moses struck the rock and produced the water, the city-dwelling Israelites assumed they had witnessed the intervention of God. What they actually saw was a practical means to reach a hidden spring by a knowledgeable desert dweller.

This incident gives a clear example of how springs so easily enter the realm of the supernatural. Ground water abides by the same laws of physics and chemistry as anything else. The difference, however, is that its behavior is not readily observable and therefore often gives the *appearance* of being supernatural. To any reasonable Israelite, an explanation of these happenings that involved fractured limestone or evaporating mineral water would have made little sense. A much more accessible explanation was that God simply produced water from a rock.

ALL OF THIS GOES A LONG WAY toward explaining why pennies still accumulate in mall fountains. Ground water and wells have a supernatural aura associated with them that comes from thousands of years of belief and practice. Even though most people don't seriously believe in water spirits anymore, the act of tossing a coin into a fountain while silently reciting a secret wish can send a superstitious shiver up the spine of even the most committed rationalist. This undercurrent of mysticism is a legacy of our collective past and is inescapable when people think about ground water and wells. But coins in fountains aren't the only manifestation of ground water mysticism.

Consider, for example, the healing springs.

ADDITIONAL READING

Bord, J., and C. Bord. 1985. *Sacred waters: Holy wells and water lore in Britain and Ireland*. London: Granada. 232 pp.

Issar, A. S. 1990. *Water shall flow from the rock: Hydrogeology and climate in the lands of the Bible*. New York: Springer. 213 pp.

Keller, W. 1982. *The Bible as history.* 2d rev. ed. New York: Bantam Books. 465 pp.

Mackinlay, J. M. 1893. *Folklore of Scottish lochs and springs.* Glasgow: William Hodge. 364 pp.

Chapter Three

Springs of Healing

Near dusk, three women approached the well. The youngest of the three, a woman of about thirty, was quietly weeping as she cradled a sickly-looking three-year-old child in her arms. One of the older women drew a cup from her pocket, dipped water from the well, and offered it to the grieving mother. The mother took the cup, put it to the child's lips, and gently urged it to drink. Meanwhile, the third women dipped a linen bandage in the well water and then wound it around the child's head, covering its eyes. Finally, they filled an earthen jar with water and reverently withdrew, quietly padding off into the gathering darkness.

This heartbreaking scene occurred in Scotland near the end of the nineteenth century and was witnessed by a gentleman named J. R. Walker, who was studying Scottish folk customs. Medical science at the time was largely helpless in the face of the epidemics that regularly swept through the countryside. For many people, particularly poor people, healing wells and springs were the only available source of protection or relief from disease.

Healing wells or springs were common in the Scottish countryside at that time and were often specific for particular ailments. In Perthshire, near the Old Castle of Garth, there were two springs referred to as Fuaran n' Gruarach and Fuaran n' Druibh, respectively. Translated, the titles mean "Well of the Measles" and "Well of the Whooping Cough." During the whooping cough epidemic of 1882, a visiting gentleman named James Mackintosh Gow described the scene when all of the local children were brought to Fuaran n' Druibh:

> It was the custom to carry the water from the well and place it in a natural cavity of a nearby boulder of mica schist, and then give the patients as much as they could take, the water being administered with a spoon made from the horn of a living cow, called a *beodhare* or living horn; this, it appears, being essential to affect a cure.

Another malady-specific well was located on the highway between Agr and Prestwick and had a reputation for curing leprosy. In fact, King Robert Bruce, who was reported to have contracted this disease, visited the well and benefited from the experience. As an offering to St. Ninian, who is credited with originally sanctifying the well, King Bruce dedicated a hospital for lepers at the site.

Pediatric wells were particularly popular in nineteenth century Scotland, which, given the rate of infant and child mortality, is not too surprising. One of these was Chapel Wells, located near the sea in Kirkmaiden Parish between the Bays of Portankill and East Tarbet. Chapel Wells is another of those wells where there is a rare eyewitness account of the procedures used and cures wrought. A Dr. Robert Trotter visited this well in 1870 and reported what he saw:

> These wells—three natural cavities in a mass of porphyritic trap rock—are within the tide mark, and are filled by the sea at high water of ordinary tides. The largest is circular, five feet in diameter at the top and five feet deep. The other wells almost touch it, and are about one foot six inches wide and deep respectively.

Dr. Trotter then reported the procedure:

> The child [who was described as "sickly"] was stripped naked, taken by one of the legs, and plunged headforemost into the big well till completely submerged: it was then pulled out, and the leg used to hold the child submerged in the middle well. Finally, the eyes were washed in the smallest well. The entire procedure was altogether like the Achilles and Styx business, only much more thorough. An offering was then left in the old chapel, on a projecting stone inside the cave behind the west door, and the cure was complete.

The folklore that preserves tales of curative springs and wells are properly called "myths." A myth is a traditional story whose specific purpose is to explain an otherwise inexplicable natural phenomenon, in this case, that water from certain wells or springs, when consumed or used for bathing, cured a variety of human ailments or the dramatically improved the symptoms. This was a natural phenomenon both puzzling and extremely important to people; as such, it required some explanation. The most readily available explanation was purely mystical: that the well or spring water had been endowed with some supernatural power, usually by a saint in the dim past.

But there are rational explanations for the apparent curative power of some wells and springs. One such explanation is the "placebo effect." Long experience has taught physicians that having a patient report dramatic improvement following treatment is no evidence that the treatment actually worked. In controlled studies of some of the most painful diseases, terminal cancers for example, simply giving patients sugar pills (that is, a placebo) and telling them that these are a new experimental treatment will result in 20 to 40 percent of the recipients feeling better.

In modern medicine, having patients "feel better" does not establish a treatment as effective therapy. What physicians look for is independent evidence that the underlying condition has been improved. But throughout most of human history, when medical science was virtually absent, having 20 to 40 percent of the ailing people feel better was a pretty good success rate. It is entirely possible that many dramatic cures popularly attributed to magic well water were, in fact, application of the placebo effect on a large scale.

But because the placebo effect was as yet unknown, and lacking any other plausible reason why well water should have medicinal properties, people developed myths that nicely explained what they were seeing. Often, in places like the British Isles, these myths might go back to beginning of Christianity or even beyond.

The myth surrounding St. Winifred's Well in Wales, for example, can be traced to the seventh century A.D. According to the story, Winifred, a maiden of noble birth, was approached by a young prince named Cradocus, who tried to seduce her. Being virtuous, however, Winifred rejected his advances. In a rage, Cradocus drew his sword and struck off her head. The head rolled down the hill and, where it came to rest, a spring burst forth. As the spring water touched the ground, the earth was wrenched open and Cradocus was swallowed alive, never to be seen again.

Happily, the episode had been witnessed by a wandering holy man, who was trying—with scant success—to convert the local pagans into good Christians. The holy man immediately recognized the miraculous nature of the new spring that had so efficiently done away with the evil Cradocus. He picked up poor Winifred's severed head, sprinkled water from the new spring on it, and reattached it to her body. Winifred, sprang to her feet, filled with the Holy Spirit, and devoted the rest of her life to spreading the Gospel.

Winifred's spring remained, however, and was henceforth famous for the cures wrought by its waters. In the eighteenth century, according to a gentleman named Mr. Pennant: "All infirmities incident to the human body, met with relief; the motive crutches, the barrows and other proofs of cures, remain near the well."

In the seventeenth, eighteenth, and nineteenth centuries, the healing powers of springs and wells were actually codified into accepted medical practice. An example of this was a book by Dr. James Gully of Malvern, England, *The Water Cure in Chronic Disease* (1846), which was widely read and quite well thought of. Gully was a careful, observant, rational physician who believed what his eyes saw, namely, that most of the "medicines routinely prescribed" at the time were "effete and inefficient, if not positively harmful."

Borrowing strongly from the tradition of water cures then in vogue on the Continent, Gully began a practice of hydropathy (as water cure therapy was termed) in Malvern. It is almost certainly not by chance that the site was originally a Benedictine Monastery long famous for the curative powers of its spring water. Gully considered himself, and certainly was, a rational physician. But he was not above taking some mystical help if he could get it. After all, Gully wanted to cure patients, not quibble about philosophy.

One good measure of the solid reputation Gully commanded was the fact that Charles Darwin, a sufferer of chronic digestive upset, was one of Gully's patients. Darwin was also a man who believed what his eyes told him and therefore entered into the therapy with a healthy dose of skepticism. It probably bears mentioning that he was forbidden to work while undergoing treatment and put on a strict diet that allowed no sugar, spices, butter, tea, bacon "or anything good," as Darwin sourly complained.

Darwin's chronic stomach ailments immediately began to improve. His treatment consisted of an early morning scrubbing with cold water and a rough towel, then drinking a tumbler of cold water, followed by a brisk, twenty-minute walk. He kept a cold compress on his stomach all day, except at the midday meal. He was wrapped in a wet blanket for an hour, with a hot water bottle on his feet. When not being treated, he was advised to walk or go horseback riding for exercise.

As far as Darwin was concerned, the "aqueous treatment" was a huge success. He felt better, his digestion improved, and he was able to do without the bottles of strange-tasting and largely ineffective medicines he had used for years.

So the question is, how much did the water part of the treatment have to do with the spectacular results? Strict diet, sensible exercise, and no work—this could be the prescription offered today for any middle-aged workaholic who seldom exercises and eats too much rich food. Indeed, it is entirely possible that simply being forbidden to work was the major factor in the success of the "aqueous treatment."

Here we are dealing with more than just the placebo effect. Clearly something made Charles Darwin feel much better, and it would be a mistake to rule out the possibility that spring water can be therapeutic for specific ailments.

It turns out that some spring waters do contain chemical compounds that confer particular medicinal properties, and these were widely used in medical practice. Physicians in the eighteenth and nineteenth century had a very limited repertory of therapeutic drugs, and anything that seemed to work for a particular disease was carefully noted. In these cases, it was common for a hospice to develop near a spring with more or less elaborate facilities for taking care of the patients.

One account of how springs were used in the practice of nineteenth-century medicine was given by a well-educated and influential physician, Dr. Thomas Linn, in his book *The Health Resorts of Europe*. Linn's book classified the mineral springs of Europe by the type of water available and by the kinds of ailments each water could treat. "Sulphur waters" (water containing dissolved hydrogen sulfide) were used to treat chronic rheumatism, gout, tuberculosis, paralysis, and chronic bronchitis. Because many sulfur springs were also warm, they were widely used for bathing. At Baden-bei-Wien, Austria, there were thirteen different springs recommended by Linn. The temperature of the waters ranged from 80 to 90 degrees Fahrenheit, and some of the baths were elaborately constructed. According to Linn: "The bath is surrounded by balconies, from which the friends of the bathers can see and talk to them during the bath—a very sociable sight."

Other classes of springs recognized by Linn included "saline," "purgative," "indifferent," "iron-arsenical," and "sulphurated," each of which had different medicinal uses. Many of these uses make considerable medical sense. For example, the high-iron waters were widely used to treat anemia, and good results were often obtained. Sulfate-bearing spring waters, which Linn classifies as "purgative" (laxative), were used to treat stomach, bowel, and bladder ailments. Interestingly, Linn sternly warns that purgative waters are contraindicated when renal (kidney) complications are present. This is a warning that any modern physician would apply to the use of diuretics and laxatives.

Some spring waters had medicinal properties because they contained trace elements needed for good health that were lacking in the average diet of the day. Perhaps the best example of a trace element is iodine. The thyroid gland, which is located in the neck and plays an integral role in controlling human metabolism, secretes thyroxine, an iodine compound. If iodine is deficient in the diet, thyroxine production becomes more difficult, and the thyroid gland enlarges, causing a goiter. In the days before iodized salt, goiters were very common and in some iodine-deficient areas were actually considered to be normal. Simple goiters are reversible if iodine is added to the diet in proper quantities. Before iodine was discovered, however, it was noticed that goiters could be prevented or cured by drinking the water of certain springs. (Needless to say, the water contained dissolved iodine.)

There are dozens of examples of how some spring waters were used to cure goiter. Near Saratoga, New York, there are more than more than a hundred springs, and the water is heavily charged with carbon dioxide. The Mohawk Indians evidently considered the waters to be medicinal, as the name Saratoga may come from an Indian word meaning "place of the medicine waters of the Great Spirit." One of the things that made Saratoga Springs attractive to people seeking water cures was that different springs had different chemical properties. Some had abundant iron and were used to treat anemia. Congress Spring was one that contained iodine and was recommended in the treatment of goiter. A physician named John Steel recorded his usual prescription for this ailment:

> The fountain which contains the largest proportion of the hydriodate of soda [iodine] should, without doubt, be selected by the invalid laboring under these afflictions. The water should be commenced in small doses, and the quantity gradually increased, as the stomach will bear it; and its use should be continued, at least, through the summer months.

Steel goes on to recommend the water to treat "nephritic" complaints (kidney failure), chronic rheumatism, chlorosis (a type of anemia), and paralysis. Interestingly, Steel warns that the water should not be used in cases of dropsy (swollen limbs) or phthisis (chronic bronchitis) and warns that "their use . . . in all acute or inflammatory diseases should be strictly prohibited."

In any case, it is clear that at least some of the uses of spring water—particularly the use of iron-bearing water to treat anemia or iodine-bearing water to treat goiter—had a solid medical basis even by modern standards. Humans being humans, however, the tendency was to stretch the medicinal use of the spring waters beyond their actual therapeutic value. Considering that in most cases there was no alternative treatment, however, it is hard to blame the physicians for trying.

Hydrotherapy, as water treatment at hot springs is often called, is one tradition of the Hidden Sea that remains as popular as ever. Virtually all large hot springs in the United States have facilities for bathing, and many have medical facilities for performing modern physical therapy. The real difference between how hydrotherapy is practiced now, and how it was practiced in the past, has to do with degree. In the past, holy wells and hot springs were often an invalid's only hope. Now, taking physical therapy at a hot spring is just one of many available options. People in large numbers still choose to soak their various aches and pains in warm spring waters. In some cases this is done with the mystical hope that a miracle cure will be wrought. More often, however, it is done simply because it feels good.

And that, after all, is what medicine is all about.

Additional Reading

Linn, T. 1893. *The health resorts of Europe*. New York: Appleton. 330 pp.

Steel, J. H. 1838. *An analysis of the mineral waters of Saratoga and Ballston*. Saratoga Springs, NY: G. M. Davison. 203 pp.

Chapter Four

McElligot's Pool

THE MYSTICAL TRADITIONS SURROUNDING GROUND WATER—which include wishing wells and healing springs—owe much to the power of human imagination. In ancient times, imagination was pretty much all people had when trying to puzzle out the strange behavior of ground water. Why were some springs hot, while others were cold? Why could a well be successfully dug in one place but not another? Human imagination has woven some remarkable mythologies over the years to answer these questions. But human imagination does not necessarily lead only to mysticism. Imagination is the starting point for rationality as well.

Consider, for example, McElligot's Pool.

MCELLIGOT'S POOL IS A CHILD'S STORY written by the irrepressible Dr. Seuss. The story revolves around a boy named Marco, who sits patiently hour after hour fishing (unsuccessfully) in a tiny pond named McElligot's Pool. Finally, a farmer comes by and tells him he is wasting his time:

> If you sat fifty years
> With your worms and your wishes,
> You'd grow a long beard
> Long before you'd catch fishes!

Marco considers this advice thoughtfully and peers into the pool. However, because he can't really be sure there *aren't* any fish, he concludes there *might* be some. After all, he reasons:

> 'Cause you never can tell
> What goes on down below!
> This pool might be bigger
> Than you or I know!

Once Marco is in the realm of what *might* be true, his imagination is free to soar:

> This might be a pool, like I've read of in books
> Connected to one of those underground brooks!
> An underground river that starts here and flows
> Right under the pasture! And then well, who knows?

Who knows, indeed? Warming to his theme and unrestrained by anything he can actually see, Marco comes to a startling conclusion:

This might be a river,
Now mightn't it be,

Connecting
McElligot's
Pool
With
the
Sea!

The more Marco thinks about this, the more plausible it seems. After all:

If such a thing could be,
They certainly would be!

So there we have it! Beginning with the single observation that he is not catching fish, Marco comes to the startling conclusion that McElligot's Pool is connected to the sea via an underground river. In fine Dr. Seuss fashion, Marco's musings become even more formidable as the story continues. The purported underground river becomes populated with eels having two heads, catfish being chased by dogfish (with collars, of course), cowfish with bells around their necks, fish with skis, fish with parachutes, and just about anything imaginable.

But there is more to this particular Dr. Seuss story than just fantasy. It turns out that the idea of holes in the seafloor connected to "underground brooks" has a long and varied history. During the period from the Roman Empire through the Middle Ages and well into the Renaissance, this idea formed the basis of serious scientific thought. And, as it happens, these ideas have a significant kernel of truth in them.

Curiously, the Bible originally suggested the holes-in-the-seafloor idea. In particular, the Ecclesiastes 1:7 states:

> All rivers run into the sea, yet the sea is not full.
> Unto the place from which the rivers come, thither they return again.

This passage implies that water is cycling out of the sea and back into rivers, but doesn't specify exactly how. In the Middle Ages, it was widely assumed that water passed through the seafloor by means of hidden channels, ascended through the earth, and was finally discharged as springs. These springs then formed the headwaters of rivers, which dutifully carried the water back to the sea.

The idea that the water in rivers came primarily from rainwater, and thus that water was cycling through the atmosphere rather than through the ground, was not widely held. Seneca (3 B.C.–65 A.D.) was particularly adamant on this point:

> You may be quite sure that it is not mere rainwater that is carried down in our greatest rivers. . . . Rainfall may cause a torrent but cannot maintain a steady flow. . . . Rains cannot produce, they can only enlarge and quicken a river.

This idea makes a bit more sense if you think of it in the context of the Mediterranean climate. In Egypt, for example, the River Nile floods every year, apparently without any rain falling at all. The rain is falling, of course, but in the African interior a couple of thousand miles away. So the connection between rivers and rain, which seems fairly straightforward to us, was not at all obvious to people of the ancient world.

Another reason that the holes-in-the-seafloor idea was popular was that it readily accounted for the presence of springs, which happen to be very common in Italy and Greece. In the case of springs, it is easy to see why people would look for a subterranean source of water. The sea was simply the most logical place to look for such large amounts of water. Also, it wasn't too hard for people to believe that water could readily move through the earth, which was generally believed to be perforated by an extensive network of caverns and grottoes.

The really tough part of this idea, however, was to come up with a mechanism for lifting the water so high above the oceans. One attempt to explain this was put forward in 1282 by a gentleman named Ristoro d'Arezzo. According to d'Arezzo, water was drawn up though the earth by the "virtue of the heavens," which apparently attracted water in the same way a magnet attracts iron.

A somewhat better attempt to explain how water could be lifted from the ocean floor was proposed in 1504 by Gregory Reisch, who attributed it to a sort of suction phenomenon:

> Within the earth we have shown there are open spaces. . . . Since there can be no such thing as a vacuum, vapors are drawn up from the earth and condensed . . . to issue into open air as springs.

This was a pretty good try. After all, alchemists knew about the phenomenon of capillarity in which water can be drawn upward in small tubes. However, capillarity is caused by the surface tension of water, has nothing to do with a "vacuum," and can only lift water a few feet, not the hundreds of feet needed for seawater to feed springs on the land.

In addition to lifting the water, there was another large problem with the holes-in-the-seafloor idea. Seawater is salty, but spring water is (most often) fresh. If the spring water was coming from the sea, where did the salt go?

A gentleman named Johann Joachim Becker (1635–1682) neatly solved both of these problems at once. According to Becker, a great fire existed at the earth's center, which vaporized the water. This had the twin effects of separating the salt from the water and lifting it to the mountainous heights. The vapor then condensed inside mountains (which were widely believed to be hollow) and turned into pure spring water. Becker specifically compared mountains to alembics, the distillation flasks commonly used by medieval alchemists.

The concepts espoused by Becker and others were codified by a Jesuit priest, Athanasius Kircher (1615–1680). Kircher had become interested in earth science when he was caught in a great earthquake while traveling in Calabria and Sicily in 1636. The earthquake caused considerable damage and, not surprisingly, gave Kircher a severe fright. His respect for the forces fermenting beneath the earth was intensified by a subsequent visit to the infamous volcano Vesuvius. Kircher resolved to examine the hidden interior of the earth and, if possible, unlock the secrets of the awesome power he had witnessed.

In 1664, after lengthy correspondence with other members of his order scattered across most of the known earth, Kircher formalized the idea of holes in the seafloor in his book *Mundus Subterraneus*. With a series of remarkable illustrations, he showed how water seeped through the ocean's floor (creating whirlpools, incidentally), entered a series of subterranean passages, was vaporized by fires under the earth, condensed as fresh water within mountains, and was finally discharged as springs, which then led to the formation of streams and rivers.

Kircher's views were soon pushed off center stage by another hypothesis for the origin of rivers. Between 1668 and 1670, a young French lawyer named Pierre Perrault got the novel idea of actually measuring, by means of rain gauges, the annual rainfall on the Seine River basin in France. Knowing the basin's area, Perrault quickly calculated that the 50 centimeters (20 inches) of rainfall amounted to about 60,750,000 cubic meters. In addition, he took the crucial step of measuring the volume of water that discharged annually from the Seine river, which turned out to be about 10,000,000 cubic meters. In other words, about six times as much water was available from rainfall as was actually carried by the river. The source of water needed to account for the river vanished overnight. Clearly, there was no need to resort to holes in the seafloor and mountainous alembics purifying seawater to explain the source of the water. It was suddenly obvious to everybody that, while water was being cycled, as required by Ecclesiastes, it was being cycled through the atmosphere, not through the depths of the earth.

This still left unsolved the problem of the origin of springs, which took a bit more time to solve. Around the year 1700, an academic named Antonio Vallisnieri took up the problem. Vallisnieri's interest was spurred by the spectacular springs and flowing wells found in northern Italy. For example, the town on Modena, located at the foot of the Alps, was famous for its artesian springs. All the residents had to do to obtain water was to sink a shaft 50 or 60 feet deep until they came to a dense shale rock. Then they drove an auger another 5 feet or so until they breached the shale. Once the shale was breached, water would gush to the surface under pressure. There were two possible mechanisms that could explain this phenomenon. If you believed Kircher, then you would expect that this water was in the process of being pumped from the sea to the top of the Alps. If you believed Perrault, on the other hand, then you would expect that the water was falling as precipitation onto the mountains, seeping into the ground, and moving toward the sea, not away from it.

Vallisnieri approached the problem in a disarmingly straightforward way. He went into the mountains and looked for evidence of either water gushing out of the mountaintops, as Kircher's theory predicted, or evidence of water seeping into the ground, as Perrault's theory predicted. Vallisnieri followed several streams to their sources without finding any evidence of water being forced out of the ground high in the Alps. However, with the help of some sharp-eyed shepherds, he was able to observe many places where alpine streams disappeared into the ground, obviously feeding a vast system of subterranean conduits. The problem was solved. The spring water of Modena originated as rainfall or snowmelt that simply seeped into the ground. Holes in the seafloor had nothing to do with it.

For all intents and purposes, Vallisnieri's results—published in 1723—buried the holes-in-the-seafloor theory for the origin of springs and rivers. It lay undisturbed and unmourned for well over two centuries until, as luck would have it, the theory was suddenly and spectacularly resurrected.

In 1977, two hundred miles northeast of the Galapagos islands, two oceanographers huddled in the research submarine Alvin discovered hot springs gushing from the seafloor. These springs turned out to be the final product of an extensive plumbing system that exists below the seafloor. At spreading centers—places in ocean basins where the earth is being pulled apart—new crust is formed from lava that spreads out over the seafloor. The basalt rock made by this process is very porous due to cracks and fissures that form as the rock cools. Seawater seeps though these porous rocks deep into the earth, where it is heated and returned to the surface as hot springs. Now to be fair, almost all of these hot springs discharge on the ocean floor, and almost all springs on land discharge water that started out as rain- or snowfall, just as Vallisnieri showed. Nevertheless, some of the hot saline springs found on land in volcanically active regions of the earth (such as Iceland) are fed by seawater that infiltrated through the bottom of the sea.

SO THERE WE HAVE IT, the triumph of Dr. Seuss, Kircher, and the human imagination. There really are holes in the seafloor connected to "underground brooks." And incidentally, some of the creatures that inhabit these springs on the floor of the sea are every bit as weird and bizarre as those dreamed up by Dr. Seuss. It just goes to show:

'Cause you never can tell
What goes on down below!
This pool might be bigger
Than you or I know!

Additional Reading

Adams, F. D. 1938. *The birth and development of the geological sciences.* Toronto: General Publishing. 506 pp.

Seuss, Dr. 1947. *McElligot's Pool.* New York: Random House. 52 pp.

Chapter Five

Hands of the Witch

It is tempting to leave the story of McElligot's Pool, Kircher's holes in the seafloor, and human imagination right there. After all, both Dr. Seuss and Athanasius Kircher managed to grasp a significant kernel of truth in the course of their imaginative wanderings. Imagination, it would seem, is a necessary first step in learning the hidden secrets of this world.

But while the process starts with human imagination, it doesn't stop there. The job of human imagination is to see what *might* be true, not necessarily what is true. The next step is to sift through all of the possibilities, and somehow identify those that might be right. Over the last three or four thousand years, humans have tirelessly invented schemes and frameworks for separating truth from untruth. These schemes—more properly referred to as "philosophies"—are as numerous and varied as the human race itself. But when it comes to understanding ground water, mysticism and rationality have the widest followings.

Rationality is the belief that careful observation and rigorous logic can lead to truth. Mysticism, on the other hand, is the belief that some truths are beyond human perception and can be discovered only through transcendent consciousness. The first thing to notice is that these philosophies are fundamentally different, with rationality accepting, and mysticism rejecting, perception as a basis for truth. The second thing to notice is that it is pointless to argue about whether one philosophy is "right" or not. As long as each philosophy adheres to its own particular rules, they are both right. It would be silly, for example, to argue that the rules of baseball are right and the rules of football wrong. Baseball and football are just different games, just as rationalism and mysticism are different philosophies.

What is very definitely "wrong," however, is to try to blend and mix the rules of the two philosophies together. This inevitably creates massive confusion and endless arguments.

Take, for example, the ancient practice of water witching.

THE PLACE COULD BE JUST ABOUT ANYWHERE in the United States. A man gripping the ends of a forked hazel branch, his palms up, his elbows tucked closely to his sides, and with an expression of extreme concentration, walks back and forth over a field. Suddenly the end of the branch, which had been pointing straight ahead, twists in his hands and arches downward toward the ground. Frowning, the man backs up, returns the branch to its former position, and slowly eases forward again. As he reaches the same place, the branch repeats its twisting movement. This is repeated several times, with the man approaching the same spot from various directions, until he is apparently satisfied. "Dig there," he says with absolute certainty, "and you'll find plenty of water about 60 feet down." The man's client, a solid no-nonsense farmer standing nearby, nods with satisfaction. A few days later, a hole is drilled, water found, and a successful well completed.

The man with the forked branch was practicing the art of water witching. The expression "water witching" is in use only in Scotland and America, and falsely implies some sort of sorcery or devil worship. In fact, it seems to have originated in Scotland were water finders used the witch elm to make their divining rods. When Scottish immigrants arrived in America, they adopted witch hazel as their material of choice, and "water witch" came into common usage in America as well as Scotland.

The water witch, also called a "diviner" or "dowser" in England, a "*Wassersucher*" (water seeker) or "*Rutenganger*" (rod walker) in Germany, or a "*sourcier*" (spring finder) in France, was practicing a tradition that is widespread in European and American culture. The first documentation of witching can be traced back to 1556 and the writings of Georgius Agricola, a physician working in the mining camps of Bohemia. Agricola's description of how the divining rod was used is virtually identical to modern practice:

> All alike grasp the forks of the twig with their hands, clinching their fists, it being necessary that the clenched fingers should be held toward the sky in order that the twig would be raised at that end where the two branches meet. They then wander hither and thither at random through mountainous regions. It is said that the moment they place their feet on a vein, the twig immediately turns and twists, and so by its action discloses the vein.

The difference here was that the Bohemian dowsers were trying to find veins of metallic ores rather than ground water. Interestingly, Martin Luther had seen fit to declare use of the rod to be in violation of the First Commandment ("Thou shalt have no other gods before me"). This happened in 1518, so the practice apparently predated Agricola, but before Luther the written record is mute, and the custom does not seem to have been widely practiced.

One reason Luther might have been bothered by the practice is that divining relied heavily on Christian imagery. This is evident from the manner in which the divining rod was treated. It was widely believed, for example, that the divining rod must be cut on St. John the Baptist's Day, and that the rod itself must be baptized. In some cases, the diviner would place his rod in the bed of a newly baptized child, and thereafter call the divining rod by that child's Christian name. Some diviners used elaborate incantations to empower the rod. One incantation, used by a seventeenth-century diviner, was as follows:

> In the name of the Father and of the Son and of the Holy Ghost, I adjure thee, Augusta Carolina, that thou tell me, so pure and true as Mary the Virgin was, how many fathoms is it from here to the ore?

Christianity was also involved in the first documented use of the divining rod to locate ground water. In 1568, Saint Teresa of Spain was offered a plot of land for the purpose of building a monastery. The only problem with the land, and probably one reason the offer was made in the first place, was that there was no water supply. This problem was soon solved, however, as recorded by Helen Colvill in her 1909 book *Saint Teresa of Spain*:

> But one day, Friar Antonio, standing in the church cloister with his friars, a twig in his hand, made the sign of the Cross with it . . . and then he said "Dig just here." They dug, and lo! A plentiful fount of water gushed forth, excellent for drinking, copious for washing, and it never ran dry.

It is interesting to speculate just how a method for finding metallic ores got transferred to finding ground water. Hard-rock miners like the Bohemians are very observant and perceptive people. After all, their livelihood depends on their being able to find ores that often give very few outward clues of their presence. To find them requires diligence, patience, and most of all, keen powers of observation. In the course of divining for ores, and digging deep into the ground to find the hoped-for riches, water was doubtless found with much more regularity than ore. Could it be that when the divining rod dipped, it was being attracted to water as well as to ore? If this was the case, why not use the divining rod when you wanted to find water?

Applying the divining rod to find ground water probably seemed natural enough to the Bohemian miners. Furthermore, because it was their business to dig shafts deep into the ground, they were in a unique position to check the results of the rod's predictions. What they found, apparently, was that the divining rod was a practically infallible guide to finding ground water. By the time our good Friar Antonio used his twig to find water for his monastery, it was a tried-and-true technique.

Plainly, water witching is a mystical practice. The witch believes in a reality (ground water) that is beyond ordinary human perception (hidden below the ground) but that can be discovered by transcendent consciousness (by some "force" funneled mysteriously through the witch to make a rod dip). And as a framework for discovering "truth" (that is, finding ground water), water witching has been remarkably successful. In the United States today, there are anywhere from 20,000 to 60,000 water witches who practice the art on a more or less regular basis. If these practitioners failed to find water more often then they succeeded, water witching would have died out centuries ago.

Prior to a hundred and fifty years ago, rational methods for locating ground water were not widely available and water witching didn't have much competition. But beginning in the 1850s, engineers and geologists began applying rationality to locating ground water. By carefully using indirect methods to observe the subsurface (that is, by drilling wells and recording the kinds of rocks encountered), and by using mathematics to describe the flow of ground water, these rationalists found that they could locate ground water very efficiently. Gradually, the rational approach (modern hydrogeology) supplanted the mystical approach (water witching) as the predominant method for finding ground water. But it did not replace it entirely. The reason water witching did not die out is simple—water witches are generally pretty successful at finding water.

The fact that water witches are successful has alternately fascinated and irritated rational hydrogeologists, who, being human, were curious about the practice, and who, being rationalists, devised various tests to figure out what was going on. One such test, devised by a philosophy professor from Boston University named Michael Martin, was particularly elegant. Martin buried four plastic garden hoses under a classroom rug, only one of which carried water at any given time. A witch was then asked to divine which of the hoses had the water in it. In forty tries, the witch was successful in finding the correct hose nine times, or just about the 25 percent "hit" rate that would be expected from chance.

The witch being tested by Martin, a gentleman named Paul Sevigny, was not at all swayed by the results of the trial. Dowsing, Sevigny explained, only works when it is used "seriously." "If I'm trying to find water for somebody, I'm almost always right," he added.

There is little doubt that Sevigny was being truthful. In Martin's experiment, the witch had a 25 percent chance to find the water-bearing hose, and he succeeded just about 25 percent of the time. When looking for ground water, however, the witch was dealing with much better odds. The fact is, if you pick just about any spot at random to drill, ground water will be found more than 90 percent of the time. So Sevigny's observation that "I'm almost always right," while perfectly accurate, is not particularly impressive to rationalists.

All of this, however, deals only with the *results* of witching—that is, whether the witch can actually find water in any reliable fashion. It doesn't say anything about *how* the dowsing rod actually works. Central to the witch's belief in his trade is the observation that a "force" present in ground water "causes" the dowsing rod to actually move in his hand. How would a rationalist explain this?

The first person to investigate this question in any detail, a question that has bedeviled both the believers and skeptics of water witching for centuries, was our good friend Athanasius Kircher. Fascinated by anything subterranean, Kircher wanted to know if the divining rod itself was actually moving. He had witnessed the rod's use, and was impressed by its apparent ability to sense hidden springs below the ground. In 1645, in the midst of his life's work cataloging what was known and unknown about the earth's interior, he resolved to test the usefulness of the divining rod. After all, if the rod really did work, it could do much to aid his studies.

Kircher went about investigating this problem with such straightforward simplicity that it is hard to argue with either the logic or the results even today. First, he procured a divining rod from an experienced practitioner and verified that the rod would dip if used properly. Next, he took the rod and suspended it gingerly with a piece of string so that any movement of the rod itself would be immediately apparent.

The rod was then carefully passed over a place where it had dipped in the diviner's hand. To Kircher's evident disappointment, the divining rod remained motionless. Clearly, whatever movement the divining rod exhibited was imparted not by the rod, and not by the mineral vein, but by the hand of the diviner.

Given that (1) water witches can reliably find ground water and (2) that the movements of their rods are not imparted by the water but by the witches themselves, what sort of explanation can be constructed to account for these observations? The mystical answer is that the rod serves as a conduit for the mysterious forces associated with ground water. But there is a rational answer to this as well.

A dowsing rod is typically a forked branch of peach or witch hazel, although the material the rod is made from seems to be unimportant. Witches use rods made out of plastic, iron, or whatever is handy. If the material a rod is made from is not important, what is? As it happens, the one shared feature of all of these practices is not the material of the rod but the way the rod is handled. Palms up, elbows tucked in, a firm grip, and, importantly, intense concentration.

The typical witch's rod is, in fact, a spring. If you take a forked branch and assume the position favored by most witches, the first thing you will find is that you automatically put tension on it. What this serves to do is to amplify any muscle movements, voluntary or not, produced by the witch. The committed cynic might believe that witches use this consciously to make their rods dip. While this may happen in some cases, another rational explanation is possible.

One of the side effects of intense concentration, psychological pressure, and nervousness is the tendency for your hands to tremble. If you ask any violinist what is the worst thing that can happen during a performance, he or she will invariably say it is "the shakes." "The shakes" are particularly deadly to violinists because the bow, which like a dowsing rod is an efficient spring, promptly amplifies each stray muscle twitch. The result can be devastating, with the bow literally bouncing over the strings and spoiling the music. The worst thing about "the shakes" is that there is nothing the violinist can do to stop them. In fact, the more you try to stop them, the worse they seem to get. They are strictly involuntary.

Take, then, the case of the water witch, his hands gripping a spring and his mind concentrating as hard as he can. At some point, the muscles in his hands and arms are going to twitch, just like a nervous violinist, and this twitch is going to cause the dowsing rod to dip and bounce—just as it causes a violin bow to jump. The water witch who claims that the rod dips involuntarily may well be telling the truth. The witch is not consciously trying make the rod dip, but, as the good Father Kircher showed, the witch is the real source of the rod's movement.

There is also a perfectly good rational explanation why ground water is found consistently wherever the rod happens to intercept a stray muscle movement, amplify it, and cause the rod to dip toward the ground. Ground water is simply too ubiquitous and too easy to find. But the experience of having the rod twist and dip, apparently with a life of its own, and finding ground water at that very spot, has got to be exhilarating. One experience like that is probably more than sufficient to set someone on the road to a lifetime of witching.

The average witch is content with his water-finding powers and doesn't seek to understand why he should have it. After all, even *trying* to figure this out by tests would mean abandoning the mystical approach. Some witches, however, have made the mistake of trying to find a rational explanation for this entirely mystical practice. One famous water witch, a gentleman named Henry Gross, envisioned great subterranean "domes" filled with water from which "veins" branched off. Because these veins radiated "energy," they could be sensed by a competent witch. In predicting the presence of such domes, however, Gross opened the door to a rational test. All anyone had to do was drill a series of holes in order to delineate the geometry of the structure. Unfortunately for Gross, no such domes have ever been found.

WATER WITCHES ARE NOT ALONE in misunderstanding the differences between mysticism and rationality. Most professional hydrologists are appalled by the practice of water witching and galled by the fact that people believe in it and use it. After all, according to them, modern hydrogeology offers many more powerful tools than the twitching of a peach branch.

But to attack the practice because it doesn't conform to specific rational tests is to miss the point. As a mystical practice, water witching cannot logically be tested using the rules of rationalism, as Mr. Gross so ignominiously demonstrated. Furthermore, to spurn water witches as fools who listen to the whisperings of fantasy is to spurn the human imagination itself. And without imagination, both hydrologists and water witches would be blinded. After all, mysticism is as much a child of human imagination as rationality is. And any parent knows that siblings squabble as a matter of course.

The wise parent, however, avoids taking sides.

ADDITIONAL READING

Ellis, J. S. 1917. *The divining rod: A history of water witching.* U.S. Geological Survey Professional Paper 416. 59 pp.

Martin, M. 1983. A new controlled dowsing experiment. *Skeptical Inquirer* 8(2): 138-140.

Roberts, K. L. 1951. *Henry Gross and his dowsing rod.* Garden City, NY: Doubleday. 310 pp.

Vogt, E. Z., and R. Hyman. 1959. *Water witching, U.S.A.* Chicago: University of Chicago Press. 248 pp.

CHAPTER SIX

Abraham's Edge

IMAGINATION IS A FINE ATTRIBUTE TO HAVE when looking for ground water. Whether you are a water witch searching for "forces" emanating from hidden pockets of water or a hydrogeologist trying to work out the hidden structure of the subsurface, the amount of information you have is generally limited. Given this limited amount of information, imagination can step in to connect the dots of the incomplete picture, and help find the hidden water.

But once ground water is located, imagination pretty much drops out of the picture. Wells, after all, are simply devices that collect ground water. Like any other device, they are conceived, designed, and built by human beings. Moreover, building and maintaining wells is a technology that has developed slowly over time. Like the wheel, wells had to be invented. And also like the wheel, people have been improving them ever since.

The reason people do this, of course, is that well technology is useful. Knowing how to dig and construct wells in the arid Middle East meant the difference between life and death for many ancient people. That statement is not conjecture, but is based on written records.

Take, for example, the story of Abraham.

> Now the Lord had said unto Abram, Get thee out of thy country, and from thy kindred, and from thy father's house, unto a land that I will show thee. (Genesis 12:1)

With these rather terse words, God ordered Abram (whose name had not been yet changed to the more familiar Abraham) to leave his home in the land of Haran and move away to an as yet undisclosed place. Haran was a city in what is now Turkey, and the land that God showed Abram was about six hundred miles to the south, a place called Canaan. Canaan was not exactly a paradise for tent dwellers like Abram's family. In fact, the best land—the fertile valley of the Jordan River and the coastal plain of Jezreel—were already firmly occupied. Furthermore, the Canaanite occupants of these lush lands were experienced warriors and had no intention of giving away their property to Abram or anybody else, God or no God.

So, as Abram entered Canaan, he was very careful:

> And he removed from thence unto a mountain on the east of Bethel, and pitched his tent, having Bethel on the west, and Hai on the east: and there he builded an altar unto the Lord. . . .
>
> And Abram journeyed, going on still toward the south. (Genesis 12:8-9)

What is interesting about this is that Abram was obviously sticking to the Judean hill country between the fertile Jordan Valley and the even more fertile coastal plain. These hills are made of limestone and are not particularly well suited for agriculture. However, the wooded hillsides and grassy clearings were just fine for herdsmen seeking to find pastures for their sheep and goats. Perfect, that is, if a reliable source of water could be found. One characteristic of limestone terrains is that surface streams are rare. This is because most rainfall quickly drains into subterranean fractures and solution cavities. If Abram and his family were going to prosper in this new and strange land, he would have to find a way to tap ground water.

It was here that Abraham, as he came to be called after establishing his covenant with God, had a particular edge over the stronger and more numerous Canaanites. Abraham knew how to dig and construct wells.

Just how Abraham came to possess this skill is a story in itself. The Semites, of whom Abraham's family was just one small clan, were a nomadic, warlike people. Around 2,000 B.C., they began to move northward out of the Arabian desert into the Fertile Crescent of the Tigris and Euphrates Valley to a land called Mesopotamia ("Land between the Rivers"), already occupied by several races of city-dwelling people.

The Bible gives only the barest hint of how extensive these migrations of Semitic people really were, or how much trouble they caused for the Mesopotamians. The archaeological record tells us a great deal more. For example, one clay tablet found in Mari (where Abram's father had settled) is evidently a message from a worried outpost commander:

> Say to my lord: This is Bannum, thy servant. Yesterday I left Mari and spent the night at Zuruban. All the Benjamites [that is, the Semites] were sending fire signals. From Samanum to Ilum-Muluk, from Ilum-Muluk to Mishlan, all the Benjamite villages in the Terqua district replied with fire signals. . . . The city guards should be strengthened and my lord should not leave the gate.

In short, the Semites made life miserable for the Mesopotamians. They robbed granaries when they could get away with it, they pastured their flocks in cultivated fields, and they watered their livestock at Mesopotamian wells. The Mesopotamians fought these incursions when they had to, but they preferred to use diplomacy whenever possible. As the Semites continued to migrate into Mesopotamia, an uneasy coexistence seems to have sprung up between the two peoples. The Mesopotamians stayed near their cities and grew their grain, and the Semites wandered on the fringes of the desert tending their flocks. As long as the Semites behaved within certain limits, their presence seems to have been tolerated.

This gave the Semites a chance to take a close look at Mesopotamian civilization, and what they saw was one of the most remarkable water supply systems in the world. Mesopotamian agriculture was based on irrigation, and water for irrigation was supplied through an intricate system of canals, locks, dams, and wells. The importance of water management to the Mesopotamians is indicated by the clay tablets found in the ruins of the cities. Many of these tablets are correspondence between the engineers who built and maintained the canals and wells, and the administrators, who were responsible for paying the bills.

Wells were particularly easy to construct in the fluvial sediments of the Tigris-Euphrates Valley. The technique consisted of digging conical holes into the ground, and lining them with stones as the holes deepened. These stones kept the holes from collapsing in on themselves, while allowing water to flow into the wells. The conical shape made it possible both to dig several feet below the water table and to store large quantities of water. Even though the water didn't flow into the wells very fast, their ability to store such large amounts of water more than made up for this lack of efficiency.

This well-building technology greatly interested the Semites. After all, much of their time was spent shuttling their flocks of sheep and goats between pastures and sources of water. If wells could be built in the vicinity of the pastures, raising sheep could be made much more efficient and productive. When Abram left Haran, he carried with him not only his flocks, kinsmen, and servants, but also the knowledge of how to build wells.

Upon entering the Judean hill country, Abraham immediately began putting this technology to use. While streams and creeks were scarce in these limestone hills, there were numerous places where ground water seeped out of the cracks and fissures in the rock. These seeps generally just created patches of marshy land that were useless as sources of water. The Canaanites had never paid any attention to them. But Abraham, with his experience in Mesopotamia, knew better. By excavating the soil around a seep, sinking a shaft downward to the rock, and lining the hole with laid stone, the patch of marshy ground could be transformed into a pool of cool, clear water. Water could then be lifted from the wells with buckets, and poured into troughs for sheep and goats to drink.

The Canaanites were aware of Abraham's arrival in the Judean hills, but his family was fairly small, and so, for about a generation, they largely ignored them. After twenty years of well building and grazing, however, the numbers of Abraham's sheep and goats—as well as the number of his family—had grown rapidly. It soon became clear to the Canaanites that these wells were important. And, being only human, they began to steal them. One of these incidents, with a Canaanite named Abimelech, is explicitly recorded in the Bible:

> And Abraham reproved Abimelech because of a well of water, which Abimelech's servants had violently taken away. (Genesis 21:25)

Abimelech, however, claimed ignorance of the deed:

> And Abimelech said, I wot not who hath done this thing: neither didst thou tell me, neither yet heard I of it, but today. (Genesis 21:26).

In any case, the two men came to an understanding, with Abraham saying:

> For these seven ewe lambs shalt thou take of my hand, that they may be a witness unto me, that I have digged this well. (Genesis 21:30)

Whether Abraham was paying a ransom to get his well back, or whether this was just a goodwill gesture is not clear. In any case, Abraham and Abimelech made a covenant, Abraham regained control of the well, and Abimelech went home.

After Abraham's death, however, the Canaanites redoubled their efforts to get rid of the newcomers, and they systematically went about plugging the wells. Isaac, Abraham's son, who was now the patriarch of his clan, doggedly went about reopening them:

> And Isaac digged again the wells of water, which they had digged in the days of Abraham, his father; for the Philistines had stopped them after the death of Abraham. . . . (Genesis 26:18)

But Isaac was not content with just what his father had owned, and strove to build new wells:

> And Isaac's servants digged in the valley, and found there a well of springing water. (Genesis 26:19)

Predictably, the Canaanites were not pleased with this.

> And the herdsmen of Gerar did strive with Isaac's herdsmen, saying, The water is ours. . . . (Genesis 26:20)

But Isaac was evidently not one to be easily discouraged.

> And they digged another well, and strove for that also. . . .
> (Genesis 26:21)

And so Isaac moved away and tried again:

> And he removed from thence, and digged another well;
> and for that they strove not: and he called the name of it
> Rehoboth, and he said, For now the Lord hath made room
> for us, and we shall be fruitful in the land. (Genesis 26:22)

It is perfectly clear from these passages that Abraham and Isaac's well-digging skill was a principal factor in establishing their family in the promised land. Abraham had a significant edge over the Canaanites when it came to finding water. And who knows? It may be that Abraham's edge was the difference between his famous success and the very real possibility of complete failure.

Additional Reading

Keller, W. 1982. *The Bible as history*. 2d ed. New York: Bantam Books. 465 pp.

Chapter Seven

Conflict in Paradise

The story of Abraham and how he used his well-digging technology to colonize the Judean hills brings up two themes that are repeated over and over in history. First, ground water—and the ability to utilize ground water—can greatly enhance economic prosperity. And second, once the economic benefits of ground water are recognized, people inevitably start to squabble over it. In Abraham's day, this conflict was pretty straightforward, with herdsmen "striving" over ownership of particular wells. In modern times, conflicts over ground water have become more complicated. What happens, for example, when some ground water users pump so much that they draw water away from their neighbors? Does that water "belong" to the neighbors, and are the heavy pumpers in effect stealing it? These questions have lead to some colorful and bitter disputes.

Consider, for example, the battle between Hilton Head Island, South Carolina, and Savannah, Georgia.

HILTON HEAD ISLAND, SOUTH CAROLINA, is one of the largest islands on the Atlantic Coast of the United States. Not only is it large, but it is also fairly high and dry, at least by the standards of barrier islands on the eastern seaboard. The highest point on the island is 15 feet above mean sea level, and while there are low-lying marshes and wetlands to be sure, there is also an abundance of well-drained land that is perfect for luxury homes and lush golf courses. This, combined with its warm, balmy climate, makes Hilton Head Island a subtropical paradise.

But the attributes of Hilton Head Island run deeper than just attractive land and a pleasant climate. The island is underlain by the Floridan aquifer, one of the largest accumulations of fresh ground water in the United States, extending from Florida (for which the aquifer is named) to Alabama, Georgia, and South Carolina. All told, about four billion gallons of water are pumped from the Floridan aquifer every year. This huge and easily available water supply has supported much of the economic growth in the Sunbelt over the last thirty years or so.

The availability of a cheap, reliable source of ground water underlying Hilton Head Island was one of the main reasons it developed into a thriving resort community. Tropical paradises require water to irrigate lawns and gardens and, more important, to keep thirsty golf courses green. The island is surrounded by water of course, but this water is too brackish for drinking or irrigation. There are some freshwater streams on the island, but they drain blackwater swamps and the water has a distinctly unpleasant taste. In short, the surface water available on Hilton Head Island was not going to support any sort of large-scale economic development.

But, to the everlasting gratitude of the island's developers, the Floridan aquifer was there waiting to be tapped. And tap it they did. In the early days of island development (the 1960s), all the developers needed to do was drill an open hole 150 feet or so into the Floridan aquifer and start pumping. The Floridan aquifer was so incredibly productive that any given well could pump as much as a million gallons of water per day. This water was put to immediate and profitable use. Golf courses sprang up around the island surrounded by mansions and gardens. Condominiums and hotels were built to accommodate the influx of guests, (the word "tourist" does not exist in the Southern vocabulary) and the island developed rapidly.

As the island developed, use of water from the Floridan aquifer exploded—from virtually nothing in 1960 to around 10 million gallons per day (MGD) in 1984—and was still climbing, at 14 MGD, in 1990. Because water wells could be drilled so cheaply and because the aquifer was so productive, this water use could be easily sustained.

But Hilton Head Island was not the only community in the area using the Floridan aquifer. In fact, Hilton Head was a relative newcomer in the neighborhood. Savannah, Georgia, just twenty miles southwest of Hilton Head, had discovered the watery lode of the Floridan aquifer in the nineteenth century. The first Floridan wells were drilled in Savannah in 1884, and the Georgians were quick to appreciate the economic potential of this resource. Soon, water pumped from the Floridan aquifer was the mainstay of domestic water supply, as well as supporting several water-intensive industries. By 1900, Savannah was using about 10 MGD of Floridan aquifer water, a pace that rose continuously throughout the century. By 1990, Savannah was drawing about 90 MGD from the Floridan aquifer.

Ninety million gallons per day, the flow of a respectable-sized river, is a lot of water. When this much water is drawn out of an aquifer, one inevitable effect is to dramatically lower water levels (that is, the heights to which water will rise in a well under its own pressure). By 1980, this pumpage had lowered the average water level in Savannah's portion of the Floridan aquifer by more than

100 feet. But this didn't particularly bother the Georgians. If the lower water levels meant it cost more to pump the water, that was no big deal, just a minor inconvenience.

But it *was* a big deal to the Hilton Head Islanders. When pumpage lowers water levels in an aquifer, the pressure loss gradually spreads away from the pumping center in the shape of a cone. By the early 1980s, Savannah's cone of depression had spread the full twenty miles from Savannah to Hilton Head Island (figure 7.1).

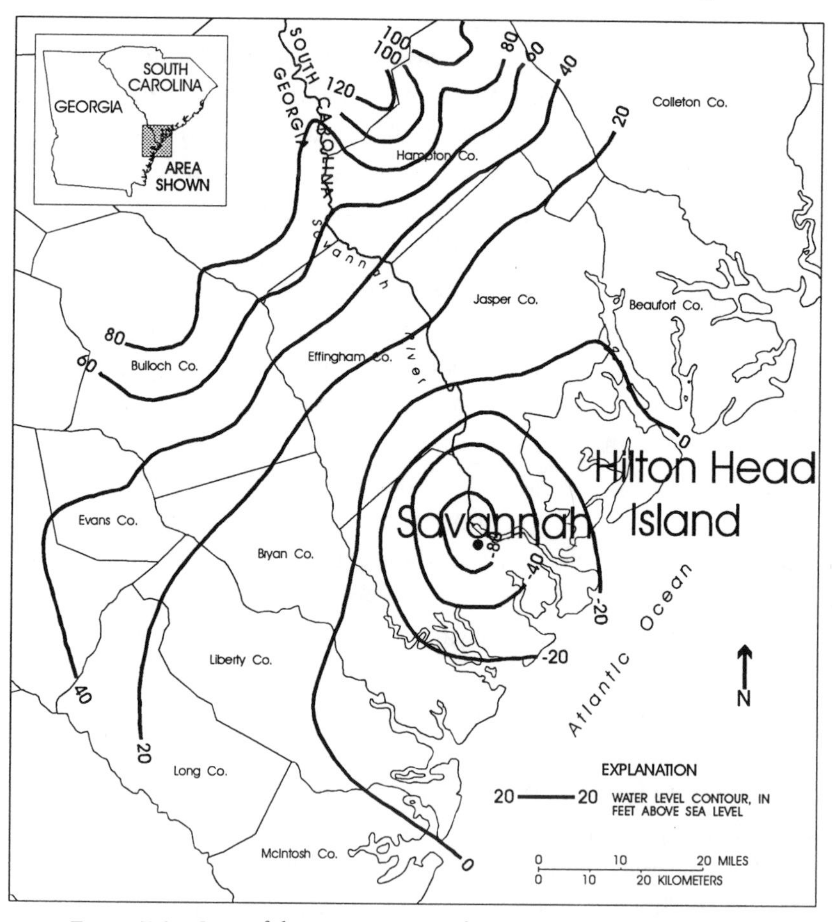

Figure 7.1 Cone of depression centered on Savannah and extending to Hilton Head Island.

This helped drop water levels on Hilton Head below the level of the estuaries just offshore, and salty water began to seep into the Floridan aquifer. If this saltwater intrusion continued, it could contaminate Hilton Head's entire water supply and render it useless—an economic disaster of major proportions.

Once it became known that salt water was encroaching on the island's sole source of water, the Hilton Head islanders loudly demanded that Savannah cut back on their pumpage. The Georgians, however, were unimpressed. They took the position that they were using their own water, not Hilton Head's, so back off. Besides, Savannah had been using Floridan aquifer water seventy years before the first well was even drilled on Hilton Head. As far as the Georgians were concerned, they had the right of prior usage.

And so the battle lines were drawn. The Hilton Head Islanders threatened lawsuits if Savannah didn't cut its pumpage. The Georgians responded by promising countersuits. There was the usual political posturing back and forth, with heated rhetoric on both sides.

Meanwhile, the salt water seeped ever closer to Hilton Head Island.

At this point, the Hilton Head Islanders had two choices. The first choice was to take Savannah to federal court and charge them with illegally encroaching on their water supply. Passions were running high on Hilton Head at the time, and suing Savannah was, to say the least, politically popular. The second choice was to study the saltwater intrusion problem and try to develop a hydrologic solution.

The problem with the lawsuit strategy was time. The law concerning prior usage rights of ground water was anything but clear, and there were no precedents for this kind of case on the books. It was entirely possible that the litigation would have to go all the way to the Supreme Court to be resolved, and that would take years. In the meantime, the salt water would be steadily encroaching. In the end, the islanders took the middle road. They kept open the option of a lawsuit, but they also decided to see a technical solution to the problem could be found.

The first order of business was to study the hydrologic problem in detail. The islanders acquired the help of various governmental agencies such as the South Carolina Water Resources Commission and the U.S. Geological Survey. They also raised, in very short order, a million dollars to support the needed studies.

WHEN THESE STUDIES GOT UNDER WAY, the first thing to do was find the saltwater-freshwater interface in the Floridan aquifer under the surrounding estuaries and see how close it really was to Hilton Head. That by itself was a major undertaking. The only way to do it was to drill a series of wells offshore of the island and sample water in the Floridan aquifer. A floating drilling rig was procured, and the entire summer of 1984 was spent drilling holes and sampling water. By moving back and forth offshore, a team of hydrologists managed to locate the wedge of encroaching salt water (figure 7.2). The news, however, was not good. The toe of the brackish water wedge under Port Royal Sound, just north of Hilton Head, was already edging under the island.

But how bad that news was depended on how fast the salt water was moving toward the island. After all, if it was moving just a few feet per year, it would take more than a hundred years for the salt water to arrive, and everybody could breathe easy. On the other hand, if the salt water were moving a few hundred feet per year, Hilton Head Island was in immediate trouble.

Because ground water systems are so inaccessible, it is usually impossible to directly measure how fast salt water moves in an aquifer. There are, however, indirect methods. Because the actual location of the salt water was known, and because its position before pumping began could be reasonably estimated, it was possible to build a mathematical model of the saltwater movement. This model could then estimate how fast the salt water was approaching the island. Even better, this model could be used to see how differ-

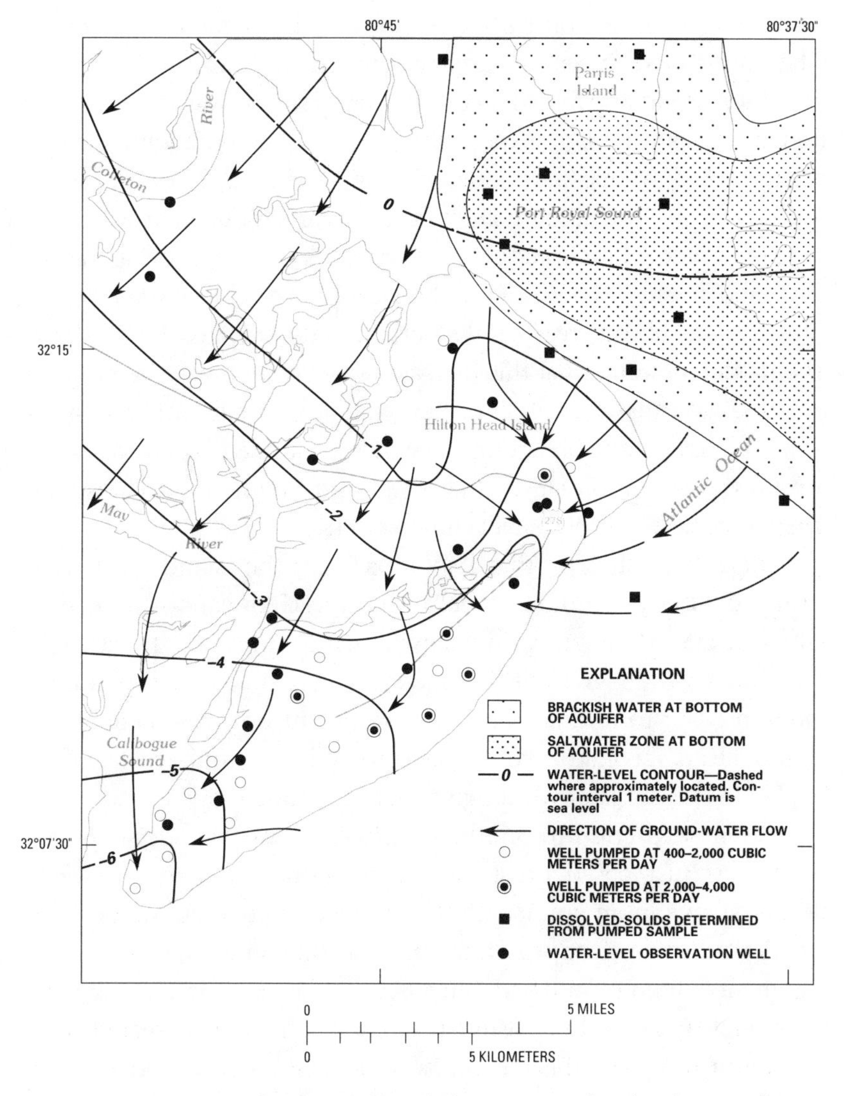

Figure 7.2 Encroachment of brackish water into the Floridan aquifer at Hilton Head Island.

ent rates of pumpage affected the rate of saltwater intrusion and to identify, it was hoped, a management strategy that would stop, or at least delay, the encroaching salt water.

This model was constructed by a quiet man named Barry Smith, who had moved to South Carolina from Indiana, and who brought with him an unflappable Midwestern calmness that was to prove very useful. For, as you might expect, the politics of the situation remained pretty heated. While the local politicians bickered, however, Smith was quietly at work, and by 1988, he had worked out an answer.

Smith's model indicated that salt water was rapidly moving toward the island at a rate of about a hundred feet per year. As of 1984, brackish water had already reached the northern end of Hilton Head Island. The model indicated that this brackish water would reach some wells by 1990, and that even if all of the pumpage on Hilton Head Island were to cease, Savannah's pumpage would continue to draw salt water toward the island. So far, all the news from Smith's model was bad. There was, however, one small glimmer of hope.

The model showed that if pumpage on the island itself were curtailed—even without any help from the Georgians—the rate of saltwater encroachment would slow down significantly. In fact, this slowdown would give the islanders about ten years of breathing room in order to find an alternative source of water. Somehow, the Hilton Head Islanders would have to reduce pumpage from 14 MGD to 9 MGD. That would give them the time they needed.

The first reaction to Smith's conclusions was skepticism. After all, it was just based on a theoretical model of how salt water *should* behave in an aquifer. In the absence of concrete proof, how did Smith know if he was right or not? But Smith had seen that argument coming, and had purposefully made a prediction with his model that could be easily checked. The model predicted that, beginning in 1990, the toe of the brackish water would reach a particular cluster of wells on the northern edge of Hilton Head Island. The U.S. Geological Survey set about monitoring those wells to see if the predictions came true. Early in 1990, brackish water showed up right on schedule. There was the proof.

Faced with the certainty of the encroaching salt water, the islanders took action. After much wailing and gnashing of teeth, water users on the island agreed to lower their collective pumpage from 14 MGD to 9 MGD, giving the islanders the time they needed to find more water. They set about finding a new source of water, from the nearby Savannah River.

All of this had an interesting effect on the Georgians. Where they had once been in solid opposition to lowering their pumpage, they now began to consider doing just that. If Hilton Head unilaterally lowered their pumpage, not to do the same could be viewed as negligent—and the specter of a lawsuit still hung over these proceedings. But also, the Savannah River was available, and it carried more than enough water to replace water production from the Floridan aquifer. The Georgians could afford to be flexible.

WHEN THE CANAANITES BEGAN TO STEAL HIS WELLS, Abraham was presented with a nasty dilemma. He could have chosen to fight over the matter, but it probably would have cost the lives of some of his family. Certainly, it would have disrupted his sheep-raising operations, and that could cause a famine. On the other hand, without his wells he couldn't raise his sheep, and that could also cause a famine. Faced with these unpleasant choices, Abraham decided to see if he could work out a compromise with the Philistines. The deal Abraham struck cost him some sheep and oxen (Genesis 21:27), but he got his wells back. Furthermore, the Philistines formally recognized Abraham's right to live in Canaan. All in all, it wasn't a bad deal, certainly better than a war he might not win.

By deciding to forgo a satisfying—if unproductive—lawsuit, and by using their resources to study and solve the saltwater intrusion problem themselves, the Hilton Head Islanders were taking pretty much the same approach Abraham took. And, not surprisingly, it worked. As long as ground water is a source of economic prosperity, arguments over who owns it and who gets to use it are inevitable. What is not inevitable, however, is that these arguments will be solved with the wisdom shown by Abraham or the Hilton Head Islanders.

But at least we know how.

Chapter Eight

Mythology or Technology

When the Hilton Head Islanders discovered they had a problem with their hitherto unlimited water supply, the approach they took to solving it was entirely rational. The islanders might be as prone as anyone to flip a penny into a wishing well, and at least a few of their wells had been located by water witches. But with hundreds of millions of dollars of real estate at risk, any thoughts of relying on mythology went right out the window. The islanders put their faith, and their money, into the technology of Barry Smith's model.

Throughout most of history, however, mathematical methods for assessing ground water availability, where it was moving, or how fast it was moving simply didn't exist. And, as is so often the case, this technology had a very humble beginning.

IN THE 1850S, a French engineer named Henry Darcy was charged with making certain improvements to the water supply for the City of Dijon. Part of the plan included using sand beds to filter and purify raw water. This seemed a simple enough task, but Darcy, being a good engineer, was uneasy. If he had been charged with building a bridge, he would have known immediately how to proceed. The art of bridge building had been practiced for centuries, and there were established designs, established materials, and established building procedures. But curiously, none of this existed for building something as simple as a filter bed to purify water.

For one thing, in order to know how large a filter bed he needed to build, Darcy needed to know how fast water would move through different kinds of sand. He soon found that there was no established engineering practice for measuring water flow through porous sands. So, being a methodical sort of fellow, Darcy built a physical model of a sand filter (figure 8.1). In this model, there were two reservoirs of water, one above and one below the sand filter. When the water level in one reservoir (measured relative to an arbitrary elevation) was higher than the other, water would flow through the filter, and Darcy could measure how fast it was flowing.

The first thing Darcy discovered was that the rate (Q) that water moved through a filter of a given cross-sectional area (A) was proportional to the water level difference (universally referred to as the "head difference," or Δh) between the reservoirs and inversely proportioned to the thickness (Δl) of the sand filter (equation 8.1). But more important, this relationship could be turned into an exact equality by adding a proportionality constant K (equation 8.2). Darcy considered the constant K to be a measure of how permeable the sand was. Later investigators discovered that K also depends on the density and viscosity of the fluid moving though the column—an important consideration if you are interested in oil rather than ground water—but Darcy was concerned only with water. This formulation of what has come to be called "Darcy's law" is still the one quoted in modern textbooks and forms the basis of modern hydrology.

$$\frac{Q}{A} \propto \frac{\Delta h}{\Delta l} \quad \textbf{(Equation 1)}$$

$$\frac{Q}{A} = K\frac{\Delta h}{\Delta l} \quad \textbf{(Equation 2)}$$

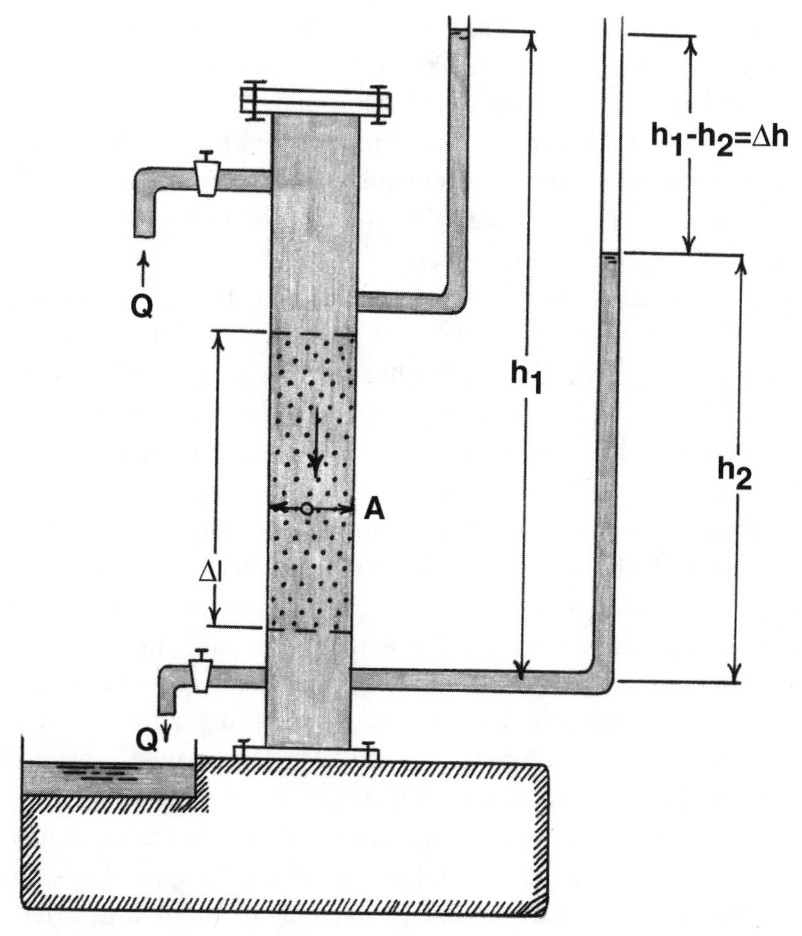

Figure 8.1 Darcy's experimental apparatus. Water was pumped through the sand filter so that the head above the filter (h_1) was greater than the head below the filter (h_2). The volume of water passing through a given area of the filter (Q/A) was found to be directly proportional to the head difference (h_1 - h_2) and inversely proportional to the filter thickness (Δl).

Darcy was thrilled with this equation because it enabled him to calculate water flow through filters of different dimensions with different h, without having to go to the trouble of building physical models with those dimensions. In essence, Darcy had replaced his *physical* model—that is, his filter apparatus—with a *mathematical* model, which was much easier to use and which could be instantly adapted to different problems. As such, it was a huge improvement in understanding how water flowed through sand.

Although Darcy was interested primarily in filter beds for purifying water, other engineers were quick to see that his model could solve other problems as well. One of the first uses of Darcy's law was for designing earthen dams. Say, for example, an engineer plans to build a dam out of soil that has a hydraulic conductivity (K) of 0.01 feet per day (10^{-2} ft/d), and wants to know how fast water will seep through his dam. In addition, say that the thickness of the dam is 10 feet, and the water level difference between the reservoir and the outflowing seep is also 10 feet. If the dam has a cross-sectional area (A) of 100 square feet (100 ft^{2}), then the only unknown in Darcy's law is Q (amount of water per unit time flowing thorough the dam). Plugging these numbers into Darcy's law (equation 8.2) gives one cubic foot of water per day (Q = 1 ft^3/day). Because there are 7.5 gallons in one cubic foot of water, the engineer can conclude that about 7.5 gallons of water per day will seep through his dam. People have been building earthen dams for thousands of years. But until Mr. Darcy came along and invented his mathematical model, they had not been able to predict how fast water would seep through them. And even though Darcy's law was derived for sand filters and widely applied to seepage through dams, it didn't take people long to realize it described the flow of ground water through natural aquifers as well.

In the eighty years that followed the formulation of Darcy's law, it was the sole mathematical model available to describe the motion of water flowing through either dams or aquifers. As useful as it was, however, it lacked something very basic. Darcy's law describes the *flow* of ground water very nicely, but had nothing at all to say about the water *stored* in the ground.

This was a big omission. Consider the plumbing of any house or building. If all the spigots and taps in the house are turned off, there is no flow of water at all. Nevertheless, the pipes are full of water, and there is a lot of water in storage. For mathematical modeling of ground water to progress, an approach was needed that would account for water storage. As often happens in applied mathematics, this new approach was provided by borrowing—some would say stealing—concepts from another discipline.

The nineteenth century was the age of the steam engine. As such, the flow and storage of heat in metals was studied in great detail by engineers and physicists. By the end of the century, physicists had developed mathematical models to describe how heat is transferred and stored in different materials. These models were used to predict how fast steam engines heated up, how hot they would get, and what sorts of systems were needed to cool them.

As it turns out, the relationship governing the flow of heat in solids is identical to Darcy's law, where the amount of heat q (calories) is equivalent to the amount of water Q (gallons), the thermal conductivity C is equivalent to permeability K, and the change of temperature with respect to distance is equivalent to the change of head with respect to distance. This being the case, could it be that equations describing the storage of heat in solids could also be applied ground water?

This question was addressed in the 1930s by an American hydrologist named Charles Theis. Taking the equation that described both the flow and storage of heat in solids, Theis simply changed the names of the variables (from "heat capacity" to "water storage"; from "temperature" to "hydraulic head"; and from "thermal conductivity" to "hydraulic conductivity"). This gave an equation, derived entirely by analogy with heat flow, describing the flow and storage of ground water.

Much to his delight, and possibly to his surprise, Theis found that, by solving this equation with some generalizing assumptions, he could accurately predict how water levels would change in a pumped well over time. Theis's equation for ground water flow revolutionized hydrology. With it, hydrologists could predict how ground water moved through an aquifer under a whole variety of conditions. They could calculate how fast water levels would decline as water was pumped from production wells, and they could predict how fast water levels would rise as water recharged an aquifer during rainfall events. All of this made it possible to calculate just how much water was available to be withdrawn from any given aquifer, and made it possible to manage water supplies much more efficiently.

One thing the ground water flow equation *didn't* do was to fundamentally change the nature of ground water, which was just as effectively hidden from view after Theis's paper was published in 1935 as it had been before. But that didn't seem to be so much of a handicap anymore. After all, now hydrologists had a mathematical model that could *calculate* ground water flow and storage, and this was almost as good as seeing it.

Things, it turned out, were going to get even better. The ground water flow equation is a partial differential equation, and is not particularly easy to solve. Theis's solution had relied on some very generalizing assumptions—many of which didn't really apply to ground water systems—and this limited its usefulness. The problem was, if these assumptions weren't made, the equation simply couldn't be solved at all. This all changed when computers became widely available. In the 1960s, hydrologists learned to use finite-difference approximations of the ground water flow equation, which computers could methodically grind through and solve. This made it possible, for the first time, to build mathematical models that were specific to certain ground water systems. Throughout the 1970s, as computer technology improved, ground water models steadily improved as well.

By the time the saltwater intrusion problem at Hilton Head Island was discovered, hydrologists had a very good repertoire of models available for simulating the behavior of ground water. The one Barry Smith used for Hilton Head could account for ground water movement, not only due to differences in head, but also due to density differences between fresh and salt water. In one sense, the Hilton Head Islanders had timed their problem perfectly. If the saltwater intrusion had started just ten years earlier, the technology for calculating how fast the salt water moved wouldn't have been available, and solving their problem would have been that much more difficult.

THROUGHOUT MOST OF HISTORY, myths were pretty much the only available tools for explaining the hidden, and often strange behavior of ground water. Why, for example, did some wells produce sweet, fresh water, while other wells nearby produced salty water? In post-Christian mythology, the most common explanation for this sort of thing was that the fresh wells had been blessed by a saint sometime in the dim past, whereas the salty wells remained under the pagan influence of the devil. It is easy to look down our technological noses at such myths, but they did provide a cogent explanation for the observed phenomenon, and such explanations did make sense in the context of people's experience. The main practical problem with the mythological explanations is that they make no useful predictions—such myths do not give any particular advice on how to deal with things like saltwater intrusion.

Like a myth, Barry Smith's model of saltwater intrusion at Hilton Head Island provided a cogent explanation for the observed phenomenon, and the explanation made sense in people's experience. But Barry's model did one other thing—it made predictions as to how the salt water would behave in the future. Even though the predictions of ground water models may not always be entirely right, they can and do give guidance to people responsible for using and protecting ground water supplies.

In a very real sense, ground water models are the technological equivalent of myths.

PART TWO

Bays of the Sea

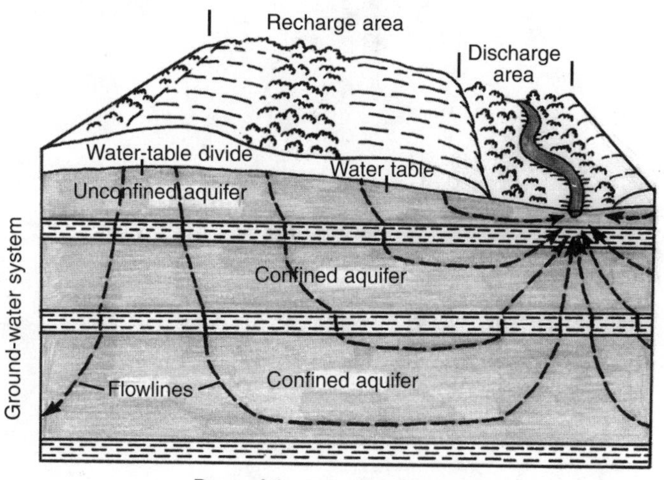

The rocks that form the crust of the earth are in few places, if anywhere, solid throughout. They contain numerous open spaces, called interstices, that hold the water found below the land, and which is recovered through springs and wells. There are many kinds of rocks, and they differ greatly in the number, size, shape, and arrangement of their interstices and hence in their properties as containers of water.

Oscar E. Meinzer, U.S. Geological Survey (1923)

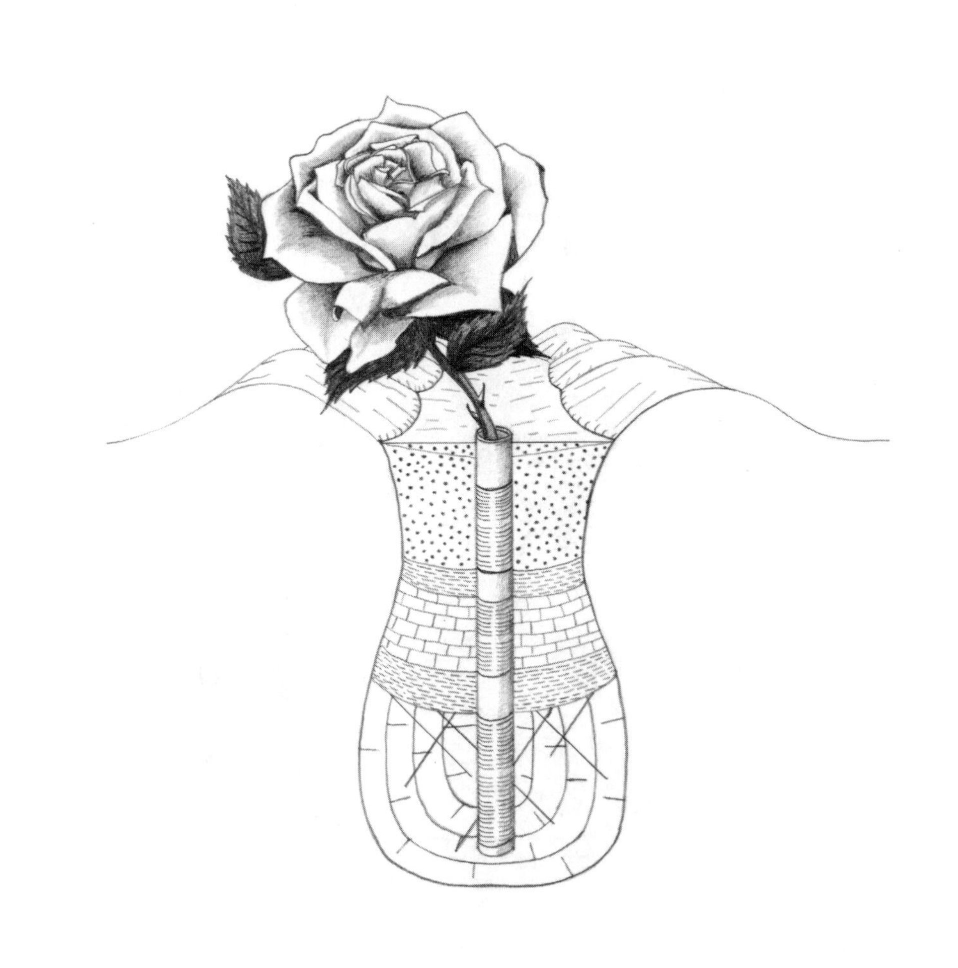

Chapter Nine

An Aquifer by Any Other Name

In western Massachusetts, the aquifer consists of crystalline metamorphic rocks that are almost completely impermeable. In places, however, these rocks are fractured and the fractures can store and transmit ground water. A 200-foot open-hole well that happens to intersect one or more fractures might produce ten or fifteen gallons per minute, which, by western Massachusetts standards, is a gusher. The individual homeowners and farmers who use these fractured crystalline rocks to supply their homes with water consider them to be a fine aquifer (figure 9.1).

In Tampa, Florida, the aquifer consists of limestones and dolomites that have extensive networks of fractures and solution cavities. A 500-foot open-hole well drilled into this aquifer will yield around 1,000 gallons per minute, and some will yield considerably more (an unusually high-yielding well—a gusher—would

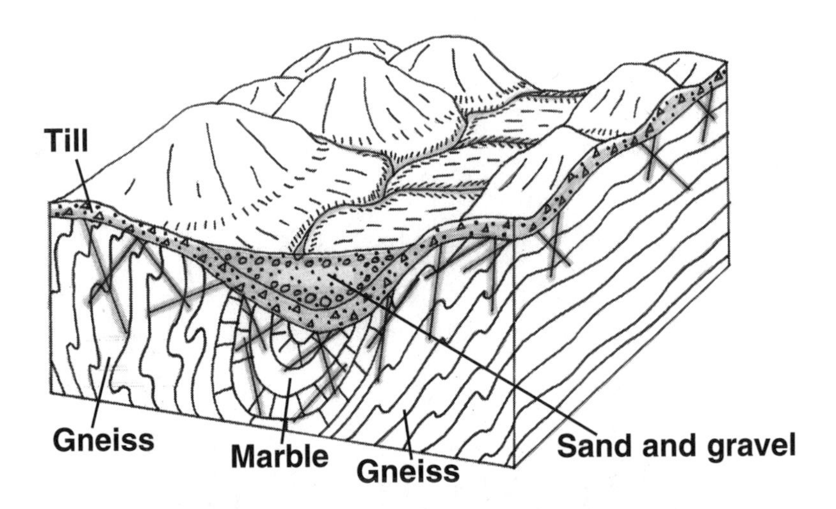

Figure 9.1
Bedrock aquifers and buried-valley aquifers characteristic of western Massachusetts (modified from Heath 1984).

have to produce more than 20,000 gallons per minute to be notable). But, by Florida standards, anything more than 500 gallons per minute is acceptable, and the local communities that use these limestones and dolomites for public water supply consider them to be a fine aquifer (figure 9.2).

In Orange County, California, the aquifer consists of alluvial sands and gravels that were deposited by rivers emptying into the Pacific Ocean. Because California is so tectonically active, these deposits have been extensively faulted. In some places, the aquifer sediments are so shallow they can be reached by wells of just a hundred feet or so deep. In other places, these sediments have been lowered by the faults and the wells may be more than a thousand feet deep. Because the sands are unconsolidated, the wells have to be cased with steel pipe and steel well screens designed to let water in while keeping the sand out. Wells constructed in this way will yield between 50 and 500 gallons per minute, and the local farmers who use these alluvial sands to irrigate their fields consider them to be a fine aquifer.

THE GEOLOGY AND HYDROLOGY of western Massachusetts, southwestern Florida, and southern California have almost nothing in common. Each area is underlain by completely different kinds of rocks, these rocks are recharged by fresh water in entirely different ways, and the technology for extracting ground water is completely different. In fact, the only thing these areas have in common is that they have an "aquifer."

The actual definition of an aquifer is simplicity itself.

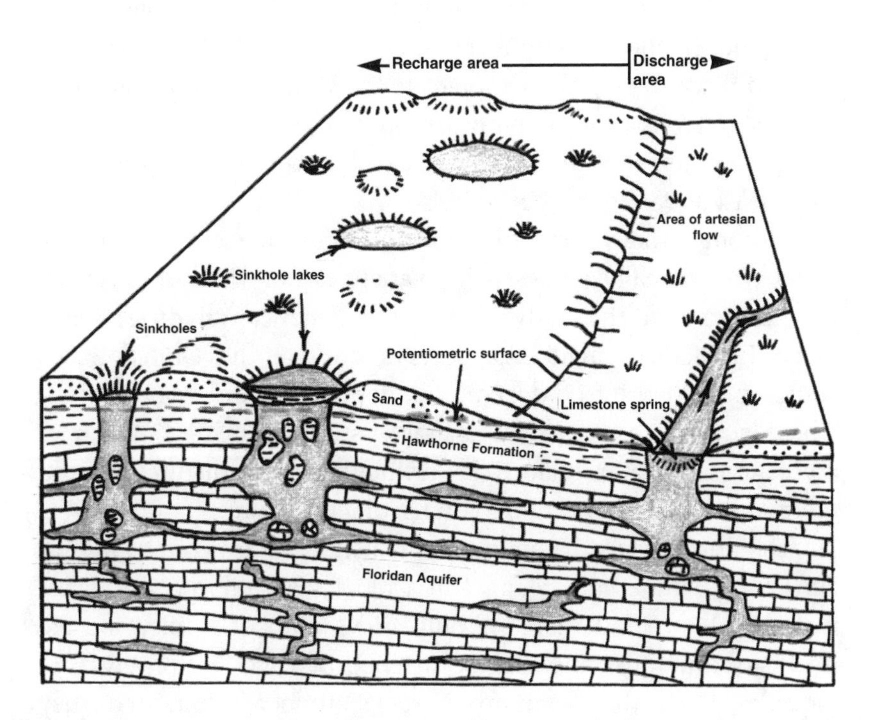

Figure 9.2 Limestone-dolomite aquifer of Florida (modified from Heath 1984).

> An *aquifer* is a geologic unit that will yield usable quantities of ground water to a well.

Note that nothing is said about what kinds of rocks qualify as aquifers, nothing is said about how shallow or deep they must be, and the only reference to the quantity of water produced is decidedly vague. "Usable quantities" of water mean very different things to different people. For a western Massachusetts homeowner, a usable quantity of water may be as low as one gallon per minute. For a southern California farmer who needs irrigation water, a usable quantity of water might be 500 gallons per minute or more.

The fact of the matter is that virtually the entire United States is underlain by an aquifer, depending of course on your definition of "usable quantities" of ground water. But the rocks that form these aquifers, their sources of water, how fast they yield this water, and the methods for recovering that water differ widely from place to place. It is easy to appreciate the variety of forms taken by surface water—lakes, swamps, mountain streams, ponds, and rivers—because they are readily observable. Ground water systems are at least as varied, maybe more so.

Although the variety of geologic circumstances surrounding aquifers is virtually endless, this variety is not without a certain organization. In the early 1900s, the American hydrogeologist Oscar E. Meinzer made the observation that the United States could be divided into a dozen or so discrete ground water regions whose aquifers have similar characteristics (figure 9.3). Each of these ground water regions is distinguished by distinctive, and more or less consistent, underlying geologic features.

The northeastern United States, which includes western Massachusetts, is underlain by ancient (500-1,000 million years old) metamorphic rocks. There are only a few kinds of aquifer you are likely to run into, and the occurrence of each is entirely predictable. Where the metamorphic rocks are locally fractured, they often contain ground water that can be recovered using open-hole

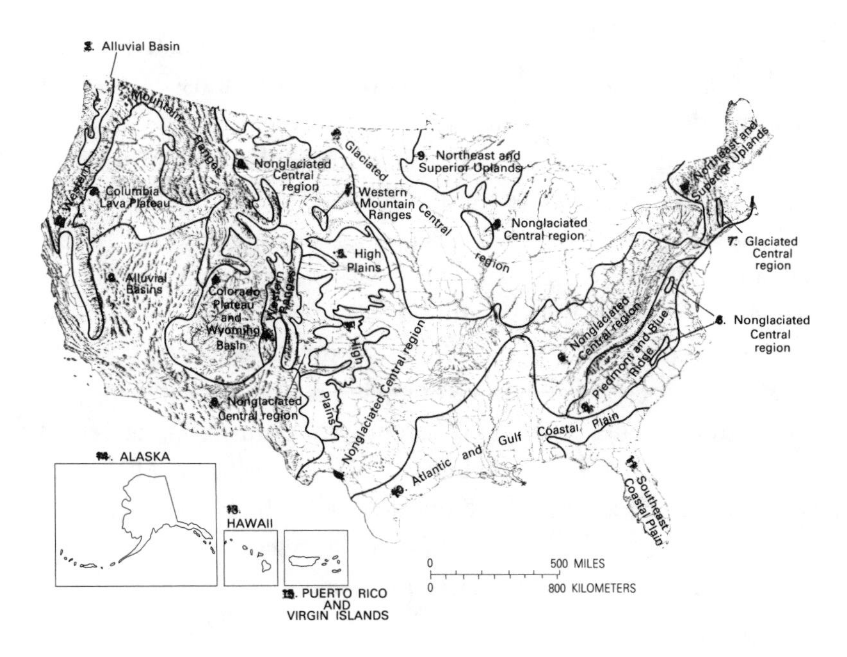

Figure 9.3 Ground water regions of the contiguous United States (modified from Heath 1984).

wells (figure 9.1). In the river valleys cut into these metamorphic rocks by water released from melting glaciers, layers of sand and gravel are often present that contain large volumes of ground water. This general area—known as the Northeast Uplands—can be neatly classified as a unique ground water region (figure 9.3).

The same kind of thing applies to Florida. Much of the southeastern United States—all of Florida and parts of Alabama, Georgia, and South Carolina—is underlain by thick accumulations of limestones and dolomites that yield large quantities of ground water. These carbonate rocks are known collectively as the Floridan aquifer, and the area underlain by the Floridan is called the Southeastern Coastal Plain region.

Finally, Southern California lies in a region dominated by alluvial basins. California is located in one of the most tectonically active parts of the world due to the collision between the North American and Pacific plates. As these huge plates grind together, blocks of rock are alternately lifted (forming mountains) or lowered (forming basins). Alluvial sediments washing off the mountain highlands fill the basins with layers of sand and gravel, which in turn store large amounts of ground water. This area has therefore come to be called the Alluvial Basin region.

Many factors contribute to the regional flavor of different parts of the United States. Climate is certainly important, as are geography, economics, and history. But one important factor often overlooked is the occurrence of ground water, which often leaves an indelible stamp on the history of particular regions of the country.

Take, for example, the Shenandoah Valley.

CHAPTER TEN

The Last Resort

THE SHENANDOAH VALLEY OF VIRGINIA, which lies within the Blue Ridge-Appalachian Valley Region, is remarkable for its physical beauty, for the fertility of its soils, and for the variety of its spring waters. Between Blacksburg and Staunton, Virginia, there are dozens of springs which, in the nineteenth century, held various reputations for miraculous healings. Their names are as colorful as their reputations: Red Sulphur, White Sulphur, Yellow Sulphur Springs were just a few. Other names are a bit more descriptive of the particular water the spring produced: Salt, Sweet, Warm, Hot, Sweet Chalebate (meaning "iron-bearing"), and Salt Sulphur Springs, among others.

These springs are unusual for a number of reasons. First of all, there is tremendous variety in the temperature and chemical composition of the spring waters, from rather hot (around 100° F) and highly mineralized to cool (around 40° F) and fresh. Second, these springs are remarkable for the reputation they earned, very early in their history, for effecting miraculous cures to numerous human ailments. Accordingly, from the 1700s to the present day, an assortment of invalids from all over the world have sought the curative powers of the springs of Virginia.

The variety of spring waters available, however, presented the nineteenth-century invalid with something of a challenge to find that particular spring whose water would be most therapeutic for the medical condition at hand. The medical community of the time could offer some limited advice. The *New and Comprehensive Gazetteer of Virginia*, published in 1835, observed:

> The White Sulphur acts, when taken in doses of two or three glasses at a time, as an alternative, exercising on the system much of salutary influence, without the evil effects of mercury,—used in larger quantities it becomes actively diaeretic [*sic*] and purgative. The Salt Sulphur is more remarkable than the White, for the latter property; but not equal to it in the former. The Red Sulphur, in addition to the qualities which it has in common with the last mentioned springs, is remarkable for its action on the pulse, which it reduces considerably in a short time. The Sweet springs . . . are of the class of waters called acidulous, and are valuable as a tonic in cases of debility, and in all varieties of dyspepsia which are unaccompanied by inflammation. The Hot springs are celebrated for their efficacy in cutaneous, rheumatic, dyspeptic and liver complaints. Dr. Bell . . . observes that all that has been performed by the Bristol, Buxton and Bath waters in England may be safely claimed . . . by the Virginia springs just enumerated.

As the reputation of the medicinal waters grew, resorts developed around the more celebrated springs. It became common for invalids to wander from resort to resort in the valley trying to find just the right kind of water for whatever ailment was troubling them. It was generally acknowledged that one should start at the White Sulphur Springs, and stay for a week or two to "improve the condition of the stomach." After the White Sulphur, however, there were no hard and fast rules. One patient might retire to the Sweet Springs to rehabilitate an overpurged digestive system. Another might go to the Salt Springs to take advantage of its iodine. Still another, particularly if troubled by arthritis, might head to the Hot Springs, where bathing in the warm water brought relief.

But there were always those sad cases for whom the spring waters could do nothing. These unfortunates wandered from one resort to another, looking in vain for relief, until either money or physical stamina gave out. It became common practice to refer to the sad individuals who had nearly run out of springs to try, as being "down to their last resort."

THE VIRGINIA SPRINGS GAINED A REPUTATION for healing powers very early in American history. In 1750, a physician named William Walker visited the Warm Springs (so named because of the constant 98° F temperature of the water) and remarked that the hospitality was excellent, but that the settlement "would be better able to support travelers was [*sic*] it not for the great number of Indian warriors that frequently take what they want from them, much to their prejudice."

Five miles south of the Warm Springs is the Hot Springs, so called because the water is somewhat hotter (around 104° F). Here Dr. Walker, in the midst of occasionally hostile bands of Indians, found six invalids that had come to attempt a cure of their various ailments. In 1761, a bathhouse was built around the Warm Springs, the first of its kind in the United States; it attracted people from all over the East Coast and soon had a high reputation for relieving various sorts of physical ailments. Such was the beginning of the resorts in the Virginia spring country.

Perhaps the most eloquent testimony to the medicinal effects of the Warm Springs was provided by a Confederate soldier named Sam Watkins. Exactly one hundred years after the bathhouse was first constructed, Watkins's First Tennessee Regiment was transported to the Shenandoah Valley by train, discharged at the railhead of Millboro, and ordered to march the twenty miles over Warm Springs Mountain to Warm Springs. According to Watkins:

> They had a large bathhouse at Warm Springs. A large pool of water arranged so that a person could go in any depth he might desire. It was a free thing, and we pitched in. We had no idea of the enervating effect it would have upon our physical systems, and as the water was but little past tepid, we stayed in a good long time. But when we came out we were limp as dishrags. About this time the assembly sounded and we were ordered to march. But we couldn't march worth a cent. There we had to stay until our systems had had sufficient recuperation.

By the 1830s, the Virginia springs were an important part of established medical practice. Interestingly, it seems that one malady for which the spring waters proved particularly useful was sterility. As Edward Pollard observed: "In sterility the Sweet Springs water is regarded as a specific." Dr. Burke from Alabama was even more blunt in his assessment: "These waters excite the animal passions, and inspire the mind with pleasurable sensations. Aged persons will find youth and vigor at the bottom of this noble fountain."

The early purveyors of the springs, being practical people, understood that romance was the first step to a successful conception. Accordingly, they arranged candlelight balls, encouraged romantic strolls, and provided comfortable nuptial accommodations. So successful were these efforts in facilitating pregnancy that young unmarried ladies actually ceased bathing in the spring waters "lest something might happen."

Indeed, so successful were the resorts in providing a pleasant, romantic environment, that by the end of the 1830s, most of people attending the Virginia springs were neither invalids nor barren wives or sterile husbands. Rather, they were members of the rising aristocracy of the South come north to escape the heat, malaria, and yellow fever that characterized southern summers. They also came to have fun, and there was plenty of fun to be had. The pleasant, cool mountain air, the blue sky, the bright summer sun, and most of all the brilliant company, contributed to the party atmosphere. At the most popular springs, there was amusement to be had by simple conversation, playing cards (whist and faro, the forerunners of modern bridge), horse racing, and dancing. Occasionally, there was even a duel.

MOST OF THE PEOPLE who made their way to the springs in the early nineteenth century were concerned mainly with having a good time. Few gave much thought to how the hot springs might have come to be or what was the source of their medicinal properties. Unlike many hot spring areas, there is not much of a mystical tradition surrounding the Virginia springs. This being the case, the rationalists have largely had their way in explaining how the springs came to be.

The rocks that underlie the Shenandoah Valley were deposited as sediments in the shallow, warm seas that surrounded North America during the early Paleozoic era (between 600 and 400 million years ago). Because the seas were warm and shallow, and because the rivers draining the continent did not carry large amounts of sand and clay, much of the sediment laid down was carbonate material, derived from the shells of various sea creatures. This led to the accumulation of thick beds of limestone (calcium carbonate) and dolomite (calcium-magnesium carbonate). Interbedded with these carbonates are beds of quartz sandstone, which record the ebb and flow of ancient beaches.

A few hundred million years after the sediments were laid down and buried, Africa collided with North America. This collision bent and folded the now-lithified (that is, no longer unconsolidated, but turned into actual rock) limestone and sandstone beds. You might think that the mountains are rocks bent upward, and the valleys rocks bent downward, but that is not the case. Some of the rocks (the quartz sandstones) are much more resistant to erosion than others (the carbonates). Thus the mountains we now see are simply more resistant to erosion and therefore stand up higher than the surrounding softer rocks. The Shenandoah Valley is a valley because it is underlain mostly by limestones and dolomites. The surrounding ridges are mountains because they are underlain by hard sandstones.

The process of bending and folding, as you might expect, causes fractures to form in the rocks that, once exposed to the surface (figure 10.1), carry water down into the earth. Most of these fractures do not penetrate particularly deep underground, and when the water is discharged, a cold spring is formed. Where the fracture system does run deep into the earth, however, the water it carries is heated by the normal geothermal gradient (that is, the deeper you go, the hotter the earth becomes). It is estimated that waters discharging from the Warm Springs and Hot Springs have circulated down about 4,800 feet, or a little less than a mile, into the earth.

As one might expect, the different flow paths taken by the shallow-circulating (cold spring) waters and the deep-circulating (hot spring) waters lead to wide variations in water chemistry. Table 10.1 shows the chemical composition of Warm Springs water, and water from a cold spring that was known as the "Montgomery White Sulphur Springs" (not to be confused with the White Sulphur Springs in West Virginia) in the nineteenth century.

The most obvious difference, of course, is the 42° F difference in temperature, which reflects the deeper flow path of the Warm Springs water. Water discharging from the Warm Springs circulates as deep as one mile below the ground, whereas the Montgomery White Sulphur Springs water circulates only a few hundred yards.

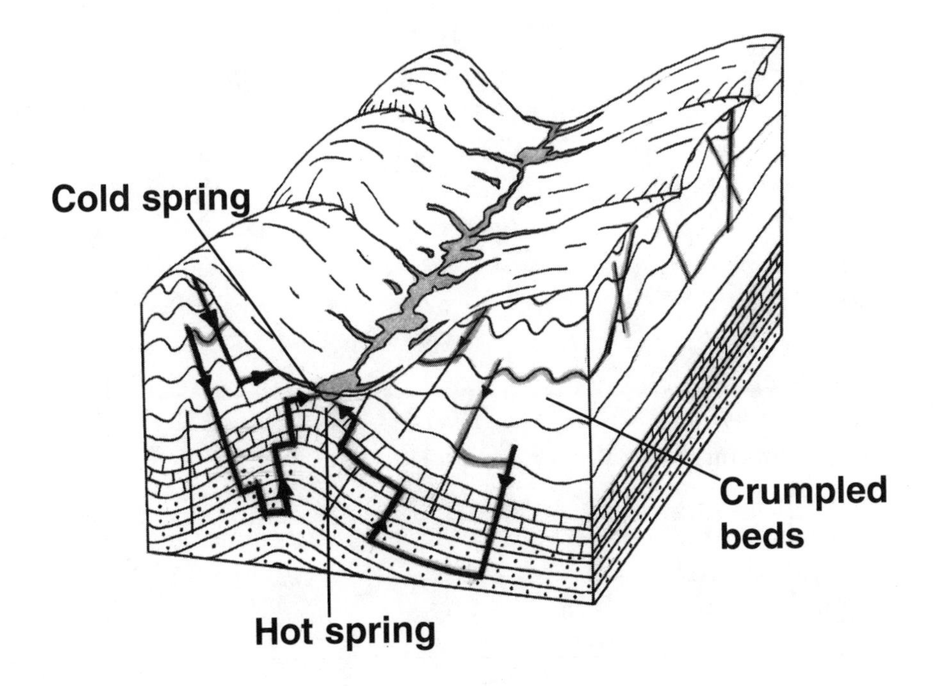

Figure 10.1
How the temperature of spring waters in the Shenandoah Valley is controlled by the depth of circulation.

There are other, more subtle differences between the water chemistry of the two springs. Note, for example, that concentrations of sodium and potassium are considerably greater in the Warm Springs water. Because sodium and potassium are progressively leached from rocks as water moves through the fractures, these greater concentrations in the Warm Springs water seem logical. Similarly, the higher concentration of sulfate in the Warm Springs water reflects progressive dissolution of the mineral gypsum (calcium sulfate) present in the limestone rock:

Table 10.1

Chemical composition of a warm and cold spring

Constituent or property	Warm Springs	Montgomery White Sulphur Springs
Temperature (^{o} F)	96	54
pH (units)	7.62	7.09
Calcium (mg/L)	110	89
Magnesium (mg/L)	23.9	36
Sodium (mg/L)	3.81	0.51
Potassium (mg/L)	10.5	0.93
Sulfate (mg/L)	254	15
Bicarbonate (mg/L)	190	435
Chloride (mg/L)	1.76	1.37
Nitrate (mg/L)	<0.5	0.5

mineral gypsum	dissolved calcium	dissolved sulfate

$$CaSO_4 \longrightarrow Ca^{++} + SO_4^{--}, \qquad (10.1)$$

and again reflects the longer and deeper flow path taken by the water.

The reasons behind other differences in water chemistry, however, are not quite so obvious. Why, for example, are concentrations of bicarbonate in the cold spring water more than double those in the warm spring water? As water moves through the rock, calcite, the main component of the limestone matrix, is progressively dissolved, producing calcium and bicarbonate ions:

limestone (calcite)	hydrogen ion	dissolved calcium	dissolved bicarbonate

$$CaCO_3 + H^{+} \longrightarrow Ca^{++} + HCO_3^{-} \qquad (10.2)$$

Because the warm water took a longer flow path, should it not have higher concentrations of bicarbonate? As it turns out, the answer is no, and for two reasons. First, as gypsum dissolves along the flow path, the dissolved Ca^{++} ions tend to combine with HCO_3^- ions to drive equation 10.2 from right to left. This results in a net *precipitation*, not dissolution, of calcite. Second, as the temperature increases, the solubility of calcite actually decreases, which again tends to precipitate calcite. The net result of calcite precipitation is a decrease in bicarbonate concentrations.

One of the more interesting aspects of these hydrological and chemical processes is that they neatly explain the different medicinal properties of the two spring waters. The cold spring yields water relatively high in calcium bicarbonate, and drinking it in quantity is like taking an Alka-Seltzer® or a couple of Tums®. This is often just what people suffering from chronic digestive upset and heartburn need for relief. The warm spring, on the other hand, yields water relatively high in calcium-magnesium sulfate. Drinking it in quantity is an efficient laxative, like taking a dose of epsom salts—just what people suffering from chronic constipation require.

Given these chemical differences between springs, the practice of traveling from resort to resort in search of the perfect medicine for whatever ailed a body is perfectly logical.

As THE SPRING RESORTS were largely filled with Southerners grown prosperous on cotton plantations, it is not surprising that the Civil War began a long decline in the resort business. Warm Springs, Virginia, as well as Hot Springs five miles away, are some of the few that still thrive. Perhaps the biggest factor in the survival of the Hot Springs resort was that the proprietors had the foresight to build a golf course around 1890. In fact, the golf course at Hot Springs is the second oldest in the United States, and is still the centerpiece of the magnificent Homestead Resort. Many other resorts, however, such as the Montgomery White Sulphur Springs, have vanished with the aristocracy that built them.

The evidence of decline at the springs, however, did not manifest itself immediately after the Civil War. When the war was over, young ladies and their chaperons again ventured to the springs. Lavish balls were planned and people again went promenading along the Lovers' Walk at the White Sulphur Springs. While there were plenty of young ladies, however, there were almost no young men at all. They had either died on the battlefield or were battling for economic survival at home in a war-shattered economy.

All of this being the case, the arrival of Northerners at the springs in the years following the Civil War was not a welcome sight. Predictably, the Southern belles refused to even acknowledge the presence of these Northern interlopers, a sentiment that was heartily returned. It was in this tense atmosphere that the springs witnessed a truly miraculous incident of healing.

The wife of General Robert E. Lee was afflicted with incurable arthritis and was in the habit of regularly retiring to the springs. The general himself suffered from heart disease and angina, an ailment that eventually killed him, and he benefited from visiting the springs as well. In 1867, two years after the war ended, General Lee and his wife arrived at the White Sulphur Springs for a rest. The general was so handsome and his bearing so dignified and courteous that he was soon the favorite of all the young ladies present. For his part, the general was happy to have the company of these young ladies since it effectively shielded him from the weighty—and tiresome—conversation of the older folks.

In the midst of General Lee's visit, the governor of Pennsylvania and a large party of guests arrived. During "the treadmill" one evening, a time devoted to strolling around the grounds and engaging in polite conversation with other visitors, the general noticed that the Southern ladies were carefully avoiding contact with their Northern counterparts and that the Northerners were returning the favor.

General Lee spoke to the young ladies who surrounded him. "Have any of you made the acquaintance of that group over there, have they been welcomed?" Silence fell over the group of usually effervescent belles, and for once a question from General Lee went unanswered.

But General Lee was not one to shy from combat. "Can no lady introduce me?" he insisted. After a pause, he gravely informed the belles around him that it was their duty to be hospitable to strangers. He turned to the young ladies and said quietly, "I shall now introduce myself and I shall be glad to present any who will accompany me." Even now, only one lady, Christiana Bond, would go with him. As they advanced toward their former enemies, Miss Bond asked the obvious question.

"But General Lee, did you never feel resentment towards the North?"

This question was asked a man who had seen his men fight, starve, and die against overwhelming odds for four years, who had asked for and received more loyalty from his men than any general since Alexander the Great, and who, after nearly beating the North on the very Pennsylvania soil from which these people came, had been run to ground and defeated at a tiny courthouse named Appomattox. This General Lee turned to the young girl and said solemnly, "I believe I may say, speaking as in the presence of God, that I have never known one moment of bitterness or resentment."

Moving across the no-man's-land that separated them, General Lee and Miss Bond greeted the Northerners and welcomed them.

From the day the Civil War ended, Robert E. Lee devoted himself entirely to the task of healing the rift between North and South. This episode at the White Sulphur Springs was only one small incident in that healing process, and one of the few that met with any success in his lifetime. The springs of Virginia, it seems, could ease not only the rheumatic complaints of invalids, but the hurts of a wounded nation as well.

Additional Reading

Reniers, Perceval. 1941. *The springs of Virginia: Life, love, and death at the waters, 1700-1900*. Chapel Hill: University of North Carolina Press. 301 pp.

LOW SODIUM
SPRING WATER

Chapter Eleven

The Heart of the Matter

THE DIFFERENT FLOW PATHS taken by circulating ground water in aquifers and the different water chemistry that results from this circulation occur to some degree in all ground water systems. In some cases, like in the Shenandoah Valley, these processes serve to create differences that are immediately noticeable (warm versus hot spring water). In other cases these water chemistry differences are less noticeable, but may still have important effects. These effects can include influencing human health.

Take, for example, the high-sodium water of the Atlantic Coastal Plain.

THE ATLANTIC COASTAL PLAIN PROVINCE, which covers much of the eastern United States (figure 9.3), is underlain by beds of sand and clay that were deposited by rivers washing out of the Appalachian highlands, and by the Atlantic Ocean. These sediments have a characteristic wedge shape, starting as a feather edge in the west, and thickening toward the east (figure 11.1A). Where permeable sandy sediments are exposed at land surface (figure 11.1B), they are recharged by rainfall. A portion of this recharge enters the deep aquifer system and flows eastward toward the ocean, where it is eventually discharged. The clay-rich beds, which are too impermeable to conduct much water, confine the flow of water to the sandy beds. These sandy coastal plain aquifers are some of the most productive sources of ground water in the United States.

But there is more to these coastal plain aquifers than immediately meets the eye. Consider, for example, the incidence of heart disease. Demographers have repeatedly noticed that some human populations living on parts of the Atlantic Coast have higher rates of heart disease than the population at large. In fact, between 1959 and 1961, some communities near the Atlantic Coast had almost twice the rate of heart disease that people had living farther inland. Now why would this be? What is it about living near the Atlantic Ocean that would give people a noticeably higher risk of developing heart disease?

One possible answer to this question has to do with the chemistry of ground water in the Atlantic Coastal Plain.

When rainwater first enters coastal plain aquifers in recharge areas, it doesn't contain much in the way of dissolved solids. In addition, what dissolved solids it does contain are mostly calcium and bicarbonate ions. But, as the water flows toward the ocean, all this changes. Concentrations of sodium and bicarbonate ions increase dramatically, and concentrations of calcium ions decrease. The Middendorf aquifer—which provides the water supply for numerous communities in South Carolina—is one of dozens of Coastal Plain aquifers that show this trend. In the recharge areas of Aiken County,

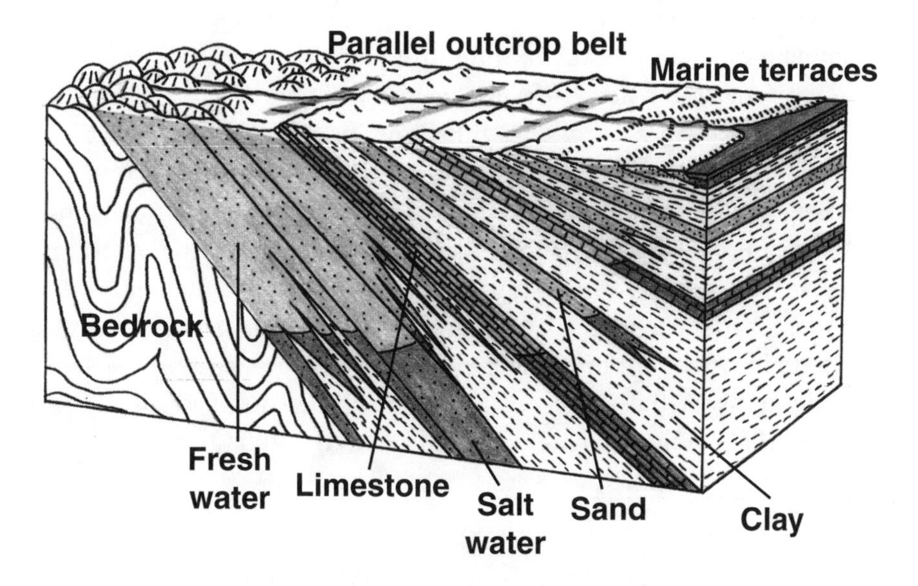

Figure 11.1A Aquifers of the Atlantic Coastal Plain.

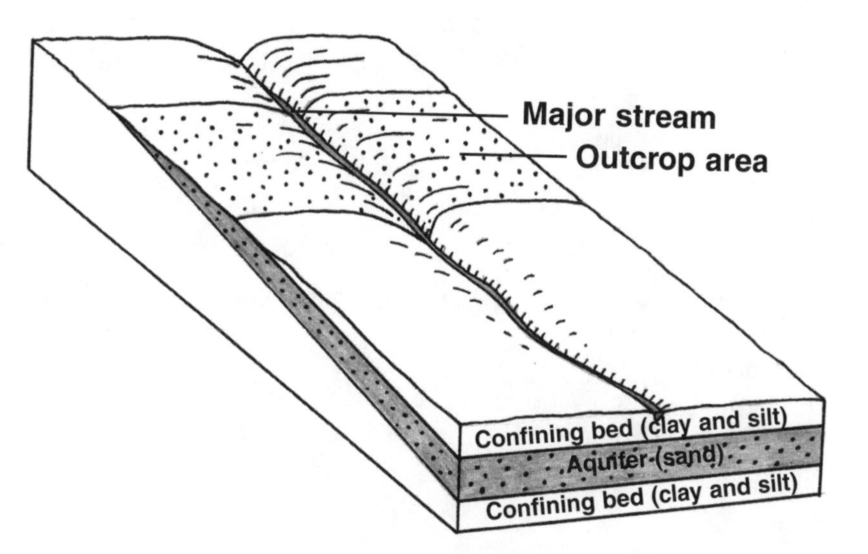

Figure 11.1B How aquifers are recharged in their outcrop areas (modified from Heath 1984).

where rainwater seeps into the aquifer, the ground water does not contain much in the way of sodium at all (table 11.1).

As the water flows toward the coast, however, concentrations of sodium increase steadily. By the time the water has reached Colleton County, about 80 miles from the recharge areas, it contains about 100 milligrams of sodium per liter (100 mg/L). By the time it reaches Charleston, more than 120 miles from the recharge areas, the ground water contains more than 500 mg/L sodium. In other words, concentrations of sodium increase more than a hundredfold as the ground water flows toward the ocean. Note that sodium is not the only constituent to increase dramatically. Concentrations of bicarbonate increase from 2 to 1,240 mg/L.

Table 11.1

Chemical composition changes as ground water flows toward the coast in the Middendorf aquifer of South Carolina

Well Location	GW Flow Distance (miles)	pH (units)	Calcium (mg/L)	Sodium (mg/L)	Chloride (mg/L0	Bicarbonate (mg/L)
Recharge area (Aiken Co.)	~5	5.1	0.8	1.3	2.0	2
Intermediate (Colleton Co.)	~80	9.0	.4	96	6.3	220
Discharge area (Charleston)	~120	8.3	3.0	540	88	1,240

High sodium concentrations in water supplies are a good news-bad news situation. The good news is that the water is beautifully soft. It is a pleasure to bath in it, and it wonderful for washing clothes and dishes. The bad news is that too much sodium in the diet increases the incidence of hypertension in humans, and hypertension is one of the leading risk factors associated with heart disease. People drinking ground water in some downgradient parts of the coastal plain take in as much as 500 mg of sodium with every liter of water they consume. Considering that the U.S. Department of Agriculture suggests a maximum daily sodium intake of 3,300 mg, this could be a health problem for some people. It is hard to be sure because so many other factors affect the health of human populations, but it is certainly possible that the presence of such high sodium concentrations in drinking water contributes to the higher incidence of heart disease found in some coastal plain communities.

But what exactly is it about coastal plain aquifers that produces these high sodium and bicarbonate concentrations in the first place? This, as it happens, is a story in itself.

DURING THE 1930S AND 1940S, D. J. Cederstrom, a geologist working with a number of coastal plain communities in Virginia that used ground water for their public water supply, was one of the first people to wonder why some of the ground water contained so much sodium and bicarbonate.

In thinking about this, Cederstrom began with assumption that the sodium was coming from interactions between ground water and whatever rocks and minerals were present in the aquifer. The aquifers consisted mostly of quartz sands, but mixed in with the sands were broken-up clam and oyster shells and bits of organic matter that are the remains of the plant and animal life that lived in the ancient rivers and seas. Also mixed in was a sprinkling of fine-grained clays and silts.

Cederstrom reasoned that quartz sand (SiO_2), which is almost completely insoluble and contains no sodium or bicarbonate ions, couldn't be generating the dissolved sodium and bicarbonate. The shell material of clams and oysters, on the other hand, was readily soluble, particularly if a source of carbon dioxide (CO_2) was available. The chemical equation describing the dissolution of calcium bicarbonate-rich shell material can be written:

shell material	water	carbon dioxide	dissolved calcium	dissolved bicarbonate

$$CaCO_3 + H_2O + CO_2 \longrightarrow Ca^{++} + 2HCO_3^{-} \quad (11.1)$$

This reaction could certainly be a source of dissolved bicarbonate, but it couldn't explain where the sodium was coming from.

But Cederstrom also knew there were traces of fine-grained clay minerals present in the aquifers. Clay minerals are composed of interlayered sheets of silica and aluminum oxides held together by electrical forces. At the edges of the clay minerals, there is often an excess of negative electrical charge. These charges attract positively charged ions (called "cations"), like sodium (Na^+) or calcium (Ca^{++}). Because sodium is so much more abundant in seawater than calcium, clays deposited in marine environments are generally coated with a layer of sodium ions held loosely in place by electrical attraction.

However, because calcium ions have a stronger positive charge than sodium—plus 2 versus plus 1—calcium ions tend to displace sodium ions on the clay particle surfaces. This takes calcium ions out of solution, immobilizes them on the clay surfaces, and puts sodium ions into solution. Furthermore, two sodium ions are put into solution for every calcium ion removed from solution. The net effect of this is to "exchange" one dissolved calcium ion for two sodium ions according to the equation

$$Ca^{++} + 2Na^{+}\text{ (clay)} \longrightarrow 2Na^{+} + Ca^{++}\text{ (clay)}. \quad (11.2)$$

Cederstrom realized that the combination of shell material dissolution (equation 11.1) with ion exchange (equation 11.2) could explain the production of ground water with high concentrations of sodium bicarbonate. The best way to test scientific theories is to see what predictions they make, and see if those predictions are actually observed. If equations 11.1 and 11.2 are added together, the resulting equation predicts that the number of sodium ions entering solution will exactly equal the number of bicarbonate ions entering solution:

$$CaCO_3 + H_2O + CO_2 + 2Na^+ \text{ (clay)} \longrightarrow 2Na^+ + Ca^{++} \text{ (clay)} + 2HCO_3^-. \qquad (11.3)$$

In other words, if you plot sodium concentrations versus bicarbonate concentrations using millimoles of ions per liter (mmol/L rather than mg/L), they should plot on a straight line with a slope of 1. And this is exactly what is observed. Figure 11.2 shows sodium concentrations versus bicarbonate concentrations for several wells tapping the Middendorf aquifer in South Carolina. It clearly fits the 1:1 ratio between sodium and bicarbonate predicted by equation 11.3.

Cederstrom, it seems, was right.

THIS PROCESS OF CATION EXCHANGE, with sodium ions replacing calcium ions, occurs naturally in virtually all coastal plain aquifers, and in many other ground water systems as well. This is a great convenience for many people because it produces the naturally soft water (without dissolved calcium) that makes washing so easy. In fact, this very same process is used by commercial water softeners to artificially convert hard (high-calcium) water into soft (high-sodium) water. The catch, of course, is that too much sodium is not particularly good for you.

Since the 1970s, when the unhealthy effects of too much sodium in the diet first became known, people have adopted a

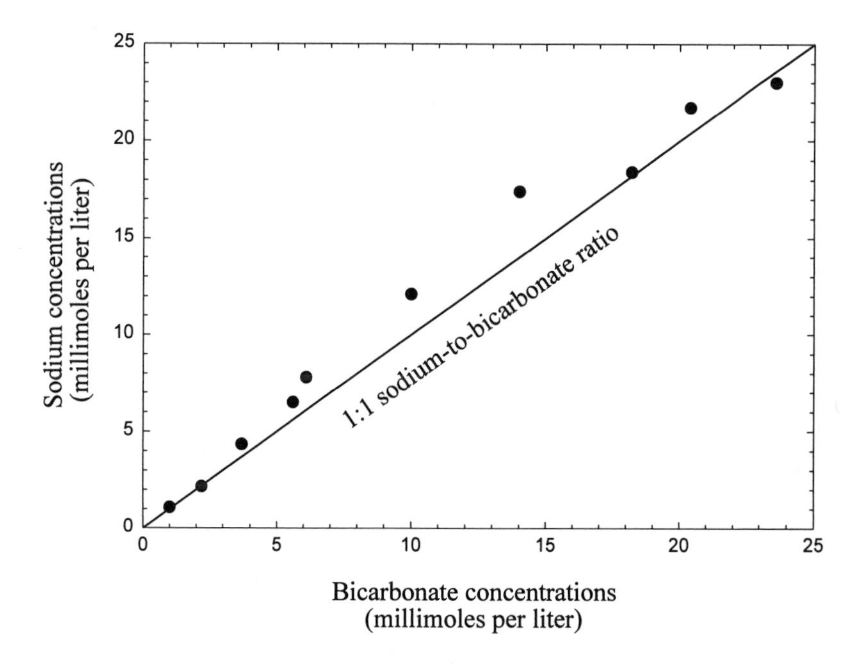

Figure 11.2 Concentrations of sodium plotted against biocarbonate for ground water from the Middendorf aquifer, South Carolina.

number of strategies for avoiding sodium in their drinking water. One popular strategy is simply to purchase low-sodium bottled water at the store and use it for drinking and cooking. In addition, many coastal plain communities with high-sodium ground water have either switched to other water sources, or have begun treating the ground water to remove sodium. In Charleston, South Carolina, for example, high-sodium water from the Middendorf aquifer is treated by reverse osmosis to remove the sodium before public distribution. This is done, in part, to make the water taste better. But it also makes it healthier to drink.

And that, after all, is the heart of the matter.

Landmeyer '96

Chapter Twelve

The Hidden Mississippi

Most aquifer systems, like those of the Atlantic Coastal Plain, are recharged directly by rainfall or by surface water from streams and rivers. In fact, most aquifer systems used for water supply are connected to some degree with adjacent bodies of surface water. The closeness of this connection, however, varies considerably from place to place. In much of the Atlantic Coastal Plain, it is barely noticeable. In other cases, the connection is so close that ground water visibly affects the behavior of surface water bodies.

Take, for example, the Mississippi River.

There are few sights as spectacular as the Mississippi River from the air. The river looks a little less enormous at 35,000 feet than it does from the ground, but not much. Its sheer size dwarfs even the huge barges that plow patiently against, or float gratefully along with, the swirling currents. Some of these barges are the size of a football field and are often linked together like sausages, six or eight at a time. And yet even these floating cities appear small against the broad expanse of the river.

Some hard numbers may help us grasp the river's size. Before its confluence with the Missouri River at St. Louis, the Mississippi is already a very large river, with an average stream flow of about 110,000 cubic feet of water per second (cfs—the standard unit used in the United States to quantify stream flow; a cubic foot of water is equal to about 7.5 gallons). The Missouri River swells the flow in the Mississippi to 183,000 cfs, and by the time the river reaches Thebes, Illinois, the average stream flow is up to 200,000 cfs. The Ohio River contributes another 281,000 cfs, and by the time the river reaches Vicksburg, Mississippi, the average stream flow is up to 579,000 cfs. At its mouth, the Mississippi River discharges an average of 593,000 cfs of fresh water to the Gulf of Mexico. To put this number into some kind of perspective, it is greater than the combined discharges of the St. Lawrence and Columbia rivers, the second and third largest rivers in North America, respectively. As impressive as the Mississippi River is by itself, there is much more to it than immediately meets the eye.

One person who watched the Mississippi River particularly closely was Samuel Clemens, better known as Mark Twain. Growing up on the banks of the river, Twain once made his living by piloting steamboats on it. But some of what he saw of the river puzzled him, and he began his book *Life on the Mississippi* with a recitation of the river's oddities:

> It is a remarkable river in this: that instead of widening toward the mouth, it grows narrower; grows narrower and deeper. From the junction of the Ohio to a point halfway to the sea, the width averages a mile in high water; thence to the sea the width steadily diminishes, until . . . above the mouth . . . it is but little over half a mile.

Considering that between the confluence with the Ohio and the mouth stream flow increases by 200,000 cfs, the fact that the river *decreases* in width by a factor of 2 does seem puzzling when you think about it. But that is not all.

> The difference in rise and fall is also remarkable. . . . The rise (and fall) is tolerably uniform down to Natchez (three hundred and sixty miles above the mouth)—about fifty feet. But at Bayou La Fourche the river rises only twenty- four feet; at New Orleans only fifteen, and just above the mouth only two and one half.

In other words, four hundred miles from the sea, river stage (the water level at a particular point) varied as much as fifty feet between periods of low water (droughts) and high water (floods). However, closer to the mouth of the river, there was progressively less difference between low and high water. Because the channel becomes narrower and more confined toward the mouth, you might expect just the opposite behavior. Where was the water going?

Another property of the river that impressed Twain was the propensity of the channel to shift and move:

> The Mississippi is remarkable in still another way—its disposition to make prodigious jumps by cutting through narrow necks of land, and thus straightening and shortening itself. More than once it has shortened itself thirty miles at a single jump! These cut-offs have had curious effects: they have thrown several river towns out into the rural districts, and built up sand-bars and forests in front of them. The town of Delta used to be three miles below Vicksburg; a recent cut-off has radically changed the position, and Delta in now two miles above Vicksburg.

The cutoffs, however, were only part of the story:

> The Mississippi does not alter its locality by cut-offs alone: it is always changing it habitat bodily—is always moving bodily sidewise. . . . Nearly the whole of that one thousand three hundred miles of old Mississippi River which La Salle floated down in his canoes, two hundred years ago, *is good solid dry ground* now [Twain's italics]. The river lies to the right of it, in places, and to the left of it in other places.

Twain had a capacity for making accurate observations without feeling the need to explain what he saw. For him, therefore, the behavior of the Mississippi River was simply remarkable. He assumed there were good reasons why the river acted as it did, but he didn't pretend to know what those reasons might be. In fact, Twain was doing what most people do today when looking out of an airliner window. He focused on the flowing surface water that could be seen and didn't consider the importance of the ground water that couldn't be seen. And, as it turns out, ground water is a critical player in the behavior of the river.

For one thing, the Mississippi River is a natural aquifer-making machine (figure 12.1). The main tributaries of the Mississippi, particularly the Missouri and the Ohio, carry enormous volumes of sediment. Furthermore, the capacity of the river to physically move all of this sediment constantly changes. At periods of high water, the velocity of the river is higher and its capacity to carry sediment is greater. At periods of low water, the river's velocity is lower and less sediment may be carried. Depending on the river stage, therefore, there is either net erosion or net deposition of sediment.

How does the river deal with these imbalances? It does so by creating meanders. Anyone who has watched a river closely knows that some parts of the channel carry water a little faster than others. As water flows around a bend in the river, a water particle on the outside of the bend moves faster than a water particle on the inside of the bend. On the outside of the bend, the faster current

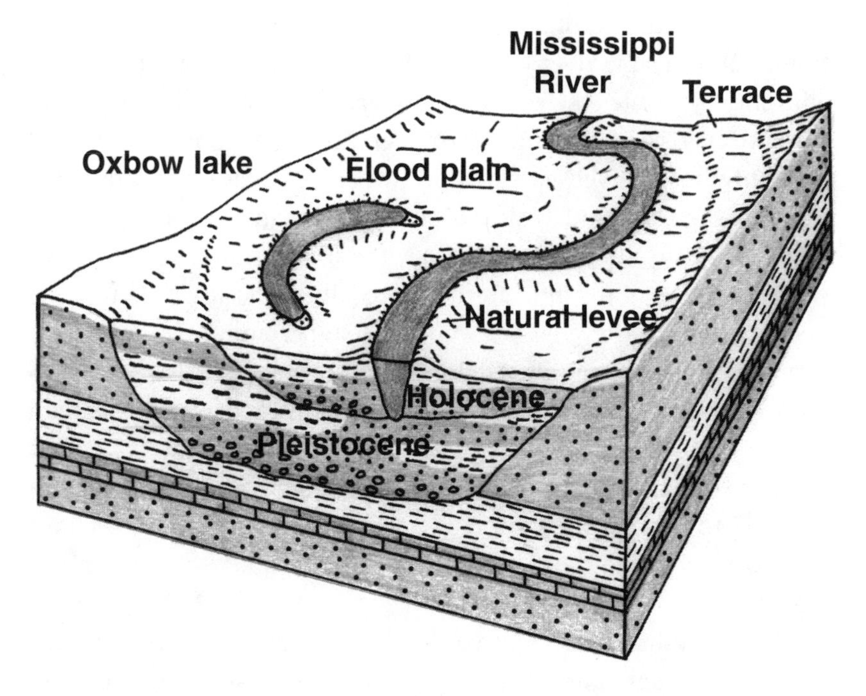

Figure 12.1 Interaction of the Mississippi River with the underlying Mississippi River alluvial aquifer (modified fromHeath 1984)

encourages sediment erosion, while on the inside of the bend, the slower current encourages sediment deposition. The combination of erosion on the outside, and deposition on the inside leads to the river shifting sideways toward the faster flowing water.

These processes lead to the dramatic shifting of the river channel that impressed Twain so much. The tendency of the river is for a bend to become increasingly exaggerated over time. This is stable for a while, until river stage rises and flows over the narrow spit of land between the loops in the river. This causes the phenomenon of "cut-offs," which periodically isolated river towns in Twain's day, and the prevention of which keeps the Corps of Engineers gainfully employed today.

So what does all of this have to do with ground water? In addition to shifting the river channel to and fro, the action of the river efficiently separates the coarser sand-sized sediments from the finer silt and clay-sized material. Coarse-grained sediments can be carried only by relatively fast-flowing water, whereas fines sediments may be carried by much slower currents. Because of this, the sediments are physically separated as the river shifts laterally. The coarser gravels and sands are deposited in the fastest water in the deepest part of the channel. As the channel continues migrating laterally, progressively finer sediments are deposited. This creates a layer-cake arrangement of clean, coarse sands and gravels overlain by fine-grained sediments. Because the sands and gravels contain virtually no silt or clay, water moves through them very easily. In addition to moving the river channel laterally, these processes have created one of the most productive ground water systems in the world.

It is of course no secret that the Mississippi River carries a huge volume of water. But consider this. Between Cairo, Illinois, and Memphis, Tennessee, the Mississippi River is about 200 miles in length. Furthermore, at normal river stage, the river averages about 2,000 feet in width and about 35 feet in depth. This immediately gives us the volume stored in the river at an average instant as 74 billion cubic feet, or about 555 billion gallons of water.

That is certainly a lot of water, but consider this also. Cairo and Memphis are about 150 miles apart on a direct line (remember how crooked the Mississippi River is), and the flood plain formed by the laterally shifting river channel averages about 70 miles wide. Furthermore, the average thickness of the water-saturated sandy sediments is about 100 feet. If we assume that the porosity of these sands is 0.3 (that is, 30 percent of the sediment volume is taken up by water between the sand grains), then we calculate that there are about 8.7 trillion cubic feet, or about 65 trillion gallons, in storage. This is about a hundred times more water than is in the river itself. In other words, between Cairo and Memphis, only about one percent of the water actually present in the combined ground water-surface water system happens to be in the Mississippi River at any particular moment. The rest, and by far the most of it, is hidden away as ground water.

This ground water may be hidden from the view of steamboat pilots or airline passengers, but it is no secret to the good citizens and farmers of Louisiana and Mississippi. In the state of Mississippi, 82 percent of all water used (except for hydroelectric power generation) is ground water and about 80 percent of this comes from the Mississippi River alluvial aquifer, or MRAA for short. In all, the MRAA produces about a billion gallons of water per day and supports much of the state's agriculture. The numbers are similar in Louisiana.

Once the enormous quantities of water stored in the MRAA are considered, some of Twain's "remarkable" features of the river begin to make sense. Take, for example, his observation that the difference between high and low water decreases toward the mouth of the river. As water levels in the river begin to rise, some of this water percolates out of the river and recharges the MRAA. In effect, the MRAA acts as a huge sponge, soaking up excess water from the river and buffering how fast it can rise.

The opposite process also occurs. During droughts, when there is little surface runoff being carried into the river, water levels in the river decrease below levels in the MRAA. This causes ground water to drain from the MRAA into the river and prevents the river levels from falling very rapidly. When one considers the vast quantities of water stored in the MRAA, it is evident that ground water recharge can keep the river flowing for a long time indeed. Near the mouth of the river, where the combined effects of ground water inflow and outflow have had the greatest cumulative effect, the rise and fall of the river become virtually nonexistent.

IT IS CONVENIENT TO MAKE A LOGICAL DISTINCTION between surface water and ground water systems. After all, surface water is out in plain view, while ground water is effectively hidden away. But there is always a connection between them. In some cases, like in the Atlantic Coastal Plain, this connection is barely noticeable. In others, like the Mississippi River and the Mississippi River alluvial aquifer, this connection can have highly visible consequences.

And this can make the Hidden Sea seem a little less hidden.

Chapter Thirteen

Basins, Ranges, and Recharge

The Mississippi River alluvial aquifer behaves in its own unique way because it is constantly receiving recharge from, or is discharging water to, the Mississippi River. All aquifer systems have sources of recharge and discharge, although few are as visible as the Mississippi River alluvial aquifer. Regardless of how apparent these sources are, however, they have a large impact on how ground water systems function, which in turn affects how much water is available for human use, and what sort of contortions humans have to go through in order to utilize it.

Take, for example, the alluvial basin aquifers of the American Southwest.

THE SOUTHWESTERN UNITED STATES is characterized by a distinctive kind of topography that geologists call "basin and range." "Range" refers to the mountain ranges that have been thrust up above the surrounding country, and "basin," to the valleys that lie between the mountain ranges. Because the Southwest is so arid, many of these basins are deserts, with annual rainfall not exceeding five inches. Clearly, finding and maintaining adequate water supplies is the key to the success or failure of human habitation in this sort of environment. And the unique hydrology of these basins and ranges plays an important role in providing these supplies.

From the air, the most striking features of basin-and-range topography are the huge fans of sediment that sweep out of the mountains and spread out onto the basins below. While rain rarely falls on the deserts in the centers of the basins (most receive less than five inches per year), it is much more common in the surrounding mountains (which receive an average of twenty-five inches or more per year). Once rainfall hits the barren rocks of the mountains, it goes right to work, washing sediment ranging in size from house-sized boulders to fine clays toward the alluvial basins. This sediment collects into great fans that spread out on the basins below.

In the process of forming alluvial fans, these masses of sediments are efficiently sorted by the moving water. At the heads of the fans, water rushes out of the mountains fast enough to carry boulder-sized chunks of rock. As the water moves out onto the fan, it slows down, and the larger sediments are deposited. Only the finest-grained sediments reach the toe of the fan. This effectively separates coarse-grained sediments from fine-grained sediments. Furthermore, the coarse-grained sediments—which are also the most permeable—are concentrated at the margins of the basins.

Rainfall and snowmelt produced in the mountains percolate into these coarse-grained sediments and efficiently recharge the underlying ground water system. As rainwater and snowmelt rush off the mountains and onto the fans, they recharge the coarse-grained, permeable sediments at the head of each fan. Because the head of the fan is considerably higher than the floor of the desert, this newly trapped ground water seeps downward under the pull of gravity toward the basin and toward the deserts (figure 13.1).

All of this is good, at least as far as constructing an efficient aquifer goes, but it gets even better. The basins and ranges are separated by geologic faults—surfaces along which blocks of the earth's crust move relative to each other. In the Southwest, the blocks of earth making up the mountains have moved upwards relative to the blocks of the basins, creating deep valleys. Over time, these valleys have gradually been filled with sediments washing out of the mountains. Thus not only are very permeable aquifers formed, but these aquifers are often one or two thousand feet thick.

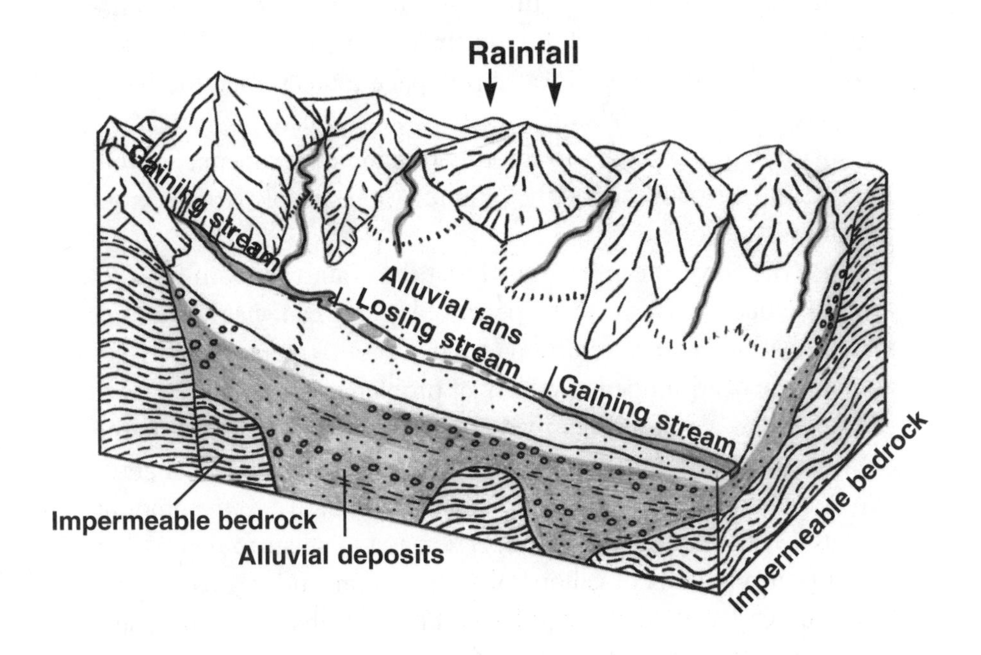

Figure 13.1 Alluvial fan aquifers of the Basin and Range ground water region (modified from Heath 1984)

Now, to be fair, not all of the sediments that fill the valleys are coarse-grained enough to be usable as aquifers. But nevertheless, the net effect is to trap water where it falls as precipitation in the mountains, and transport it to the center of the desert basins. This water is then stored in an immensely large underground reservoir.

The huge stores of ground water created by these processes are economically very important. The soils of these basins are often quite good for agriculture, and the climate is well suited for some crops. Cotton, for example, loves the hot dry climate and will thrive in these arid basins if it is given plenty of irrigation. Thanks to the unique hydrology of the Southwest—with the water storage capacity of the basins effectively combined with the recharge capacity of the mountains—ground water is readily available.

This ground water has supported the growth of a thriving agricultural economy, which in turn has fueled the growth of entire cities such as Phoenix and Tucson. In the Salt River Valley, where Phoenix is located, ground water pumpage totals something like a billion gallons per year. Much of this water is used for irrigation, but a goodly portion goes to the thirsty people of Phoenix as well. In Tucson, virtually all of the public water supply comes from ground water alone.

In some ways, it can be argued that these alluvial basin aquifers are too efficient. They are so permeable, and can yield so much water, that it is very tempting to overpump them. And when aquifers are overpumped, a variety of problems can ensue.

Any time a well is pumped, the hydrostatic pressure in the aquifer is lowered, and this causes the aquifer to compact slightly. When the amount of water pumped is small relative to the amount of water in the aquifer, this compaction is too small to detect. But if you pump billions of gallons of water a year out of a basin, and lower water levels in the aquifer several hundred feet, the compaction can be significant. Although it occurs in three dimensions, compaction is noticeable to humans in only one—the surface of the earth subsides.

It is easy enough to calculate about how much land subsidence is caused by pumping ground water. The basic relationship is that compaction (C) is proportional ($\propto$) to the aquifer's thickness (b) and the decrease in water level (Δh):

$$C \propto b(\Delta h).$$

This can be turned into an equation by adding a proportionality constant k that represents the compressibility of the aquifer. This gives us an equation,

$$C = kb(\Delta h),$$

which can be used to calculate how much land will subside due to ground water pumping.

Most sandy aquifers have a compressibility (k) on the order of 0.00001 per foot (10^{-5} ft^{-1}). Thus, if a 50-foot-thick aquifer experiences a water level decline of 10 feet due to pumping ground water, the amount of total compaction (C) will be

$$C = 0.00001/\text{ft} \times 50 \text{ ft} \times 10 \text{ ft} = 0.005 \text{ ft} = 0.06 \text{ inches},$$

which is far too small to be noticed or measured. However, if a 2,000-foot-thick aquifer—such as the alluvial basin aquifers of the Southwest—is pumped so that the water level declines 250 feet, the compaction will be

$$C = 0.00001/\text{ft} \times 2{,}000 \text{ ft} \times 250 \text{ ft} = 5 \text{ feet},$$

and five feet of land subsidence will definitely get people's attention.

Land subsidence due to ground water pumpage has been a problem in the southwestern alluvial basin aquifers since the early part of this century. Near Phoenix, for example, water levels in the alluvial aquifer declined almost four hundred feet between 1923 and 1977. Just how much land subsidence occurred due to such heavy pumping is not accurately known. But it is generally agreed that the land surface subsided between one and five feet in some places. This caused earth fissures to open in places and damaged the foundations of buildings. The alternative, however, was to simply do without water. The people of Phoenix decided that land subsidence, while a nuisance, was better than doing without water. The pumping continued.

But Phoenix, unlike Tucson to the south, is not entirely dependent upon ground water for water supply. The Roosevelt Dam, which controls stream flow in the Salt River, stores water for municipal use as well as for agricultural use. When water levels in the Roosevelt Reservoir get too high, as sometimes happens when rainfall in the mountains is unusually heavy, water is released to the otherwise dry riverbed of the Salt River. A portion of this water seeps into the ground, providing another source of recharge to the aquifer. This has served to slow the precipitous ground water level declines in the Salt River Basin. Presently, the withdrawal of ground water from the Salt River alluvial aquifer is roughly balanced by recharge from the valley margins and by recharge from the Salt River. This in turn has largely halted the land subsidence.

THERE ARE MANY OBVIOUS DIFFERENCES between ground water and surface water bodies. From the standpoint of supplying water for humans and agricultural crops, however, the most important differences have to do with the flow and storage of water. Surface water bodies *store* relatively small amounts of water (about 2 percent of all the water in the United States), whereas ground water bodies store much more (the other 98 percent). With respect to *flow*, however, the situation is just the reverse. If you were to add up all of the water discharging into the oceans at any particular moment, you would find that the majority of it is surface water, with ground water lagging far behind. The bottom line is that surface water bodies have relatively little storage but lots of flow, whereas ground water bodies have lots of storage but not much flow.

This basic difference has a huge impact on the water supply of places like Phoenix and Tucson. The basin-and-range aquifers store such huge amounts of water that it is very tempting to pump them far faster than the ground water flow can recharge them. When this is done on a short-term basis, it is no big deal. However, when this basic imbalance continues for years, the net result is precipitous ground water level declines (as the aquifer is drained) and, inevitably, land subsidence.

The whole trick, therefore, is to balance the amount of ground water withdrawn from an aquifer with the amount of available recharge. Huge amounts of recharge are supplied to these alluvial basin aquifers by precipitation in the surrounding mountain ranges. Additional recharge is supplied by leakage from rivers such as the Salt River in Phoenix or the Santa Cruz River in Tucson. But the amount of this recharge is essentially fixed, which means the amount of available ground water is also fixed. As these cities grow, the only available option for managing their water supply is to divert water from one usage to another. It is probable that, over the years, progressively more water will be diverted from agricultural use to municipal use in the Southwest. That is going to cause economic hardship for some people, but there really isn't any other alternative.

It is just a matter of recharge.

Chapter Fourteen

The Sorrows of Job

It is certainly true that the alluvial basin aquifers of the southwestern United States have had problems caused by overpumping. When too much ground water is pumped from an aquifer, the lowered water levels greatly increase pumping costs. Also, as Phoenix found out, they can also cause land subsidence. But one ground water system in the United States that has probably suffered the most from the effects of overpumping is the San Joaquin Valley of California. This aquifer system has, over the years, endured the hydrologic equivalent of the sorrows of Job.

The aquifers of the San Joaquin Valley were formed by sediments washed out of the Sierra Nevada to the east and from the Coastal Ranges to the west. The Central Valley of California is a vast trough that has been steadily sinking, relative to the surrounding highlands, for the past 120 million years. The valley has therefore served as a natural sediment trap, and has accumulated as much as 40,000 feet of sands, silts, and clays. Sediments from the Sierra Nevada, especially sediments that have washed down in the last 100,000 years or so, are predominantly coarse sands, which make an excellent aquifer (figure 14.1).

This aquifer was discovered by the farmers of the San Joaquin Valley in the last half of the nineteenth century. From the time the Spanish first came to California, it was known that the soils of the valley were exceptionally fertile. The problem was always water. Most of the valley is, by any definition, a desert, and annual rainfall seldom exceeds ten inches. Producing marketable crops in this climate requires around forty inches per year of rainfall or irrigation. During the 1800s, an extensive network of canals was used to divert water from the San Joaquin and other rivers. Digging canals, however, was difficult and expensive work. After the discovery of the underlying ground water system—and the development of drilling techniques that could reach this water—agriculture in the San Joaquin Valley literally bloomed.

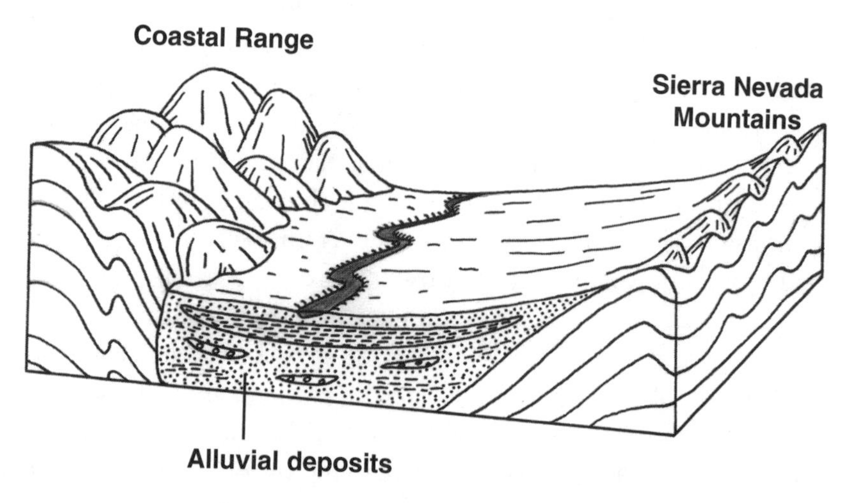

Figure 14.1 Alluvial aquifers of the San Joaquin Valley of California.

The early farmers found, to their delight, that not only was it possible to drill wells anywhere and obtain water, but there was virtually an inexhaustible supply. Many of the early wells flowed under artesian pressure, and the sands were so permeable that the average well was easily capable of yielding up to a million gallons of water per day (MGD). Once the water supply problem had been solved, the fertility of the soils did the rest. By the early twentieth century, the San Joaquin Valley was the most productive agricultural area on earth.

But it took vast amounts of ground water to keep the crops growing. By 1912, farmers were withdrawing an incredible 223 MGD of water from the San Joaquin Valley alone and an additional 100 MGD from the Sacramento Valley farther north. It was about then that the strain of this excessive use first began to show on the aquifer. For one thing, water levels in the aquifer dropped rapidly and kept on dropping (in some areas, by as much as 400 feet), which in turn started a curious chain of events.

The aquifers underlying the San Joaquin Valley consist of interbedded sands and clays. At first, of course, most of the ground water produced came out of the sands. As pumpage continued, however, the sands were literally drained and water began to be squeezed out of the clays. And as the water was squeezed out, the clays began to compact.

Land subsidence due to all of this pumpage began to be noticeable around 1920, but it was not too serious a problem. With the end of World War II, however, the worldwide demand for agricultural products boomed, and the use of ground water skyrocketed. As pumpage increased, land subsidence due to dewatering and compacting of the clays accelerated. An incredible twenty-nine feet of subsidence has been actually measured in parts of the San Joaquin Valley. This has either collapsed the wells or, in a few cases, left the well casings dangling ten or fifteen feet above the ground. In addition, the foundations of buildings were cracked, aqueducts and canals were damaged, and general havoc reigned.

The other problem was that the water from the compacting clays was nonrenewable. Once squeezed out, it was gone and could not be replaced. It soon became evident that if agriculture was to continue on the same scale as before, sources of water other than ground water needed to be found.

There were other sources of water to be found, they were just expensive to tap. Northern California has a fairly wet climate and has proportionally more runoff available to feed streams and rivers. Prior to the 1960s, most of this water simply ran off into the Pacific Ocean. With the construction of an intricate and incredibly expensive water-capturing system—a scheme that culminated in the construction of the California Aqueduct—much of this water was diverted southward through the San Joaquin Valley.

Soon, imported surface water replaced ground water as the principal source of irrigation. This, of course, was designed to take the stress off of the overpumped aquifers and halt the precipitous declines in water levels. And it worked. Unfortunately, this imported water soon proved to be as much a stress on the aquifer as had been the heavy pumping.

The ground water system of the San Joaquin Valley consists of many different sandy aquifers separated by discontinuous layers of relatively impermeable clay (figure 14.1). The shallowest water table aquifer was not generally as productive as deeper zones, and had not been as heavily pumped. The problem was that as imported surface water seeped into the ground, it recharged the shallow water table aquifer, causing water levels to rise precipitously in many areas. Also, this infiltrating irrigation water began to dissolve minerals and salts present in the soil. Soon, the shallow ground water in many places was so near land surface, and so saline, that it began to poison the roots of the very crops the irrigation was designed to nourish.

But farmers are a hard lot to discourage. If the rising ground water levels and the salts in the ground water were going to be a problem, then they would deal with that, too. It soon became the custom to install drains under irrigated fields designed to capture excess ground water and keep water levels in the shallow aquifer from rising too high. The water drained from these fields was diverted to the rivers and reservoirs in the valley.

At this point, if geologic circumstances had been different, the story might have ended happily. But it was not to be. It happens that the rocks making up the Coastal Range Mountains have an unfortunate peculiarity—they contain unusually high concentrations of selenium. The soils derived from these rocks, therefore, also have a good bit of selenium. Under natural (dry) conditions, this would be fairly innocuous. With all of the irrigation, however, the selenium began to leach out of the soil and into the ground water. The drains collected this selenium-contaminated water and delivered it directly to the low-lying wetlands of the valley.

For a good while, all of this went on largely unnoticed. In 1983, however, thousands of waterfowl began to die mysteriously at the Kesterson Reservoir. After a frantic series of ecologic and hydrologic studies, it became clear that the birds were dying from selenium poisoning.

So, after an incredible chain of events, lasting over the better part of a century, we are left with a ground water system that has been used and abused in just about every possible way. The initial overpumping led to huge water level declines in the deep aquifer—the result being massive land subsidence. To fix that problem, surface water was imported—the result being massive salt and selenium contamination of the shallow aquifer. If ever an aquifer system has suffered the sorrows of Job, this one is it.

BUT ALL IS NOT NECESSARILY LOST. The shallow aquifer system does have a certain capacity to cleanse itself of selenium contamination. Like many metals, selenium can exist in several different oxidation states. Elemental selenium exists by itself with no oxygen atoms attached and has the chemical formula Se^0. But selenium also forms stable compounds with oxygen, one being selenite (SeO_3^{--}), and another being selenate (SeO_4^{--}). As oxygen atoms are added to selenium, the compounds become progressively more soluble. Elemental selenium (Se^0) is virtually insoluble, selenite (SeO_3^{--}) is slightly soluble, and selenate (SeO_4^{--}) is very soluble.

These differences in solubility created the problem in the first place. The irrigation water percolating into the ground carried dissolved oxygen with it, and this oxygen reacted with elemental selenium in the sediments to form selenate:

$$Se^0 + 2O_2 \longrightarrow SeO_4^{--}.$$

It was this soluble selenate that was leaking into the reservoir and killing the waterfowl.

But because selenate is such an oxidized compound, it has other properties as well. We human beings are restricted to using oxygen in our respiration. Microbes living in soils and in aquifers, however, are not quite so limited and can use the oxygen in selenate for respiration. When these microbes respire selenate, they convert it back to insoluble metallic selenium:

$$SeO_4^{--} \longrightarrow Se^0 + 2O_2.$$

The selenium promptly precipitates out of solution, naturally renewing the contaminated ground water.

The net effect of all of this is that the selenium-contaminated ground water is not an unsolvable problem. If the irrigation practices are modified to minimize production of soluble selenate, microorganisms naturally present in the aquifers will, over time, improve the chemical quality of the contaminated ground water.

FROM THE VERY BEGINNING of ground water development in the San Joaquin Valley, human understanding has lagged one step behind the inflexible realities governing the aquifer system. Solving the initial agricultural problem (the lack of water) caused another problem (land subsidence). Solving the land subsidence problem (importing surface water) caused a different problem (selenate contamination of ground water). At this stage, it is difficult to say if this vicious cycle has been broken. New problems, such as salt accumulation in shallow ground water, are presently a real threat to agricultural production in the valley and will have to be faced.

But if, in our glorious ignorance, we couldn't quite give this much-abused aquifer system a knockout punch, there is some hope that it will fare better in the future.

CHAPTER FIFTEEN

An Exercise in Optimism

THE SIGNIFICANCE OF DISASTERS associated with overpumping ground water, such as had been experienced in the San Joaquin Valley, has not been lost on people living in other parts of the country. Overusing a ground water system—like overusing any natural resource—can lead to problems that most thinking people would rather avoid. The simplistic solution—and sometimes the best solution—is to just make sure that the systems are not overutilized in the first place. But life is not always that simple. In some cases, the tremendous wealth created by overutilizing a ground water system can be used to solve the very problems that go with the utilization.

Take, for example, the Ogallala aquifer

UNDERLYING MUCH OF THE HIGH PLAINS OF THE UNITED STATES (figure 9.3), the Ogallala aquifer was formed when the Rocky Mountains were thrust upward by tectonic forces. The lifting of the Rockies caused large amounts of sand and gravel to wash out of the newly created highlands. These sediments were spread out onto the High Plains by rivers and streams, and eventually accumulated to thicknesses of more than a thousand feet in places. A good deal of this sediment was scraped off by later erosion and by the advancing continental glaciers, but significant accumulations of sediment, averaging two hundred feet thick, remain beneath eastern Colorado, Nebraska, Kansas, Oklahoma, and Texas. Over the millennia, these sediments soaked up water like a sponge, and the Ogallala aquifer was born (figure 15.1).

The location of the Ogallala aquifer is, of course, just a chance occurrence. But it is doubtful that even careful planning could have placed it in a better place for agricultural purposes. The soils of the High Plains are very fertile, but much of the region lacks

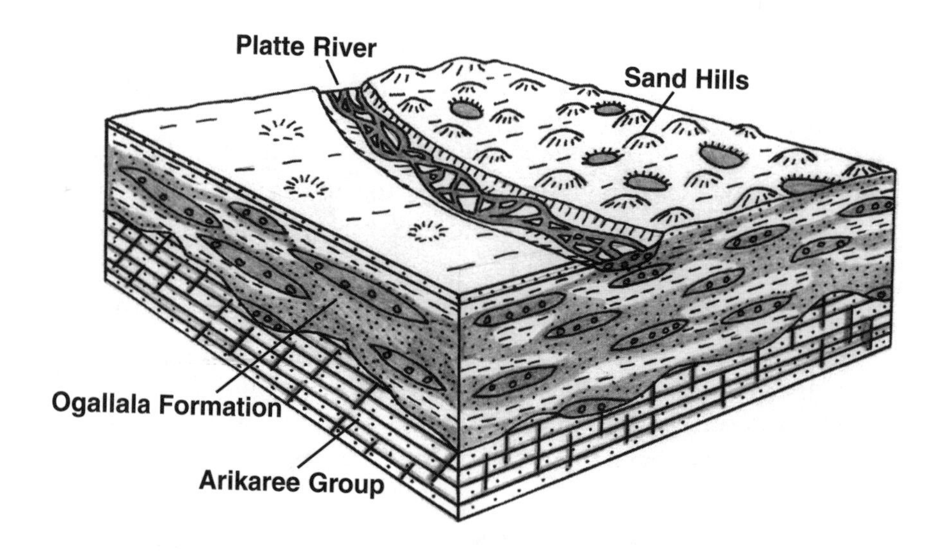

Figure 15.1 Ogallala aquifer (modified from Heath 1984)

enough rainfall to consistently grow water-loving crops like corn, sorghum, or alfalfa. When irrigated, however, these brown lands can be transformed into veritable gardens. The one or so quadrillion (million billion) gallons of water stored in the Ogallala aquifer is perfectly placed to make this transformation possible.

It didn't take the settlers who began moving into this area after the Civil War long to find out about the water stored in the Ogallala. For one thing, numerous springs issued from the aquifer. For another, it was relatively easy to hand-dig wells into the unconsolidated sediments. In fact, one of the largest hand-dug wells in the world—known appropriately as the "Big Well"—was dug in Greenburg, Kansas, in 1887 to supply water for steam locomotives. Curiously, however, it took a hundred years, the catastrophic drought of the 1930s, and the development of center pivot irrigation systems before the Ogallala aquifer was seriously exploited.

The easiest way to irrigate a field is by flood irrigation—water is simply pumped into ditches surrounding the field and diverted right to the furrows. As the name implies, this method effectively floods the field. All you need (beside abundant water) is for the fields to be properly graded so that water can flow more or less evenly over the crops. If there are swales and dips and ridges, flood irrigation will simply produce a series of ponds and islands. The rolling topography of the High Plains renders widespread application of flood irrigation impractical.

Center pivot irrigation systems, which rotate around a central point on huge wheels and sprinkle water onto crops, are perfect for this kind of rolling topography. As this technology became available in the 1950s, there was a revolution of agriculture in the High Plains states. Beginning first in the parched lands of Texas and Oklahoma, the use of center pivot irrigation spread northward throughout the 1960s. The principal driving force behind this revolution were the farmers themselves. When one farmer with a new center pivot system was able to double or triple his yield, neighboring farmers were quick to follow.

This was a very interesting time for the few hydrologists who had studied the Ogallala. Even as early as the 1960s, for example, it was known that much of the water contained in the Ogallala represented recharge that had occurred over thousands of years, and that much of the water pumped out could never be replaced. To have really large-scale pumpage meant that, sooner or later, the resource would become seriously depleted. This was understood from the very beginning. However, it was also understood that the water stored in the Ogallala could create a tremendous amount of agricultural wealth. One technical paper by the U.S. Geological Survey, dated 1965, states:

> Ground water beneath the High Plains . . . is one of the great potential economic assets awaiting development. The only certain prediction that can be made is that this asset will be developed and that this development will result in declining water levels.

That was it in a nutshell. If people wanted the economic benefits of irrigation, then there was going to be some depletion of the resource. In some cases where depletion of ground water resources occurs—such as in the San Joaquin Valley—it comes as a surprise to the people involved. Not so in the case of the Ogallala. It was known right up front that this would happen.

The collective decision, made by thousands of farmers, bankers, and politicians, was that the development should go ahead. And go ahead it did. Between 1950 and 1980, agricultural pumpage in the High Plains grew by a factor of 10—from 651 billion gallons of water per year to 7.5 trillion gallons of water per year. That, incidentally, translates to a flow of about 32,000 cubic feet per second (cfs), which is just about the flow of the Missouri River at Kansas City.

This development, however, produced a tremendous amount of wealth for the entire country. The grain produced in the High

Plains, for example, is used to feed the cattle that account for 40 percent of the nation's output of beef. If the Ogallala aquifer had never been created by the rising Rocky Mountains, if it had never been thoroughly saturated, or if it had never been developed by the farmers, that beef would never have been produced. Even more, dairy cattle and poultry farms throughout the country rely to some degree on the grain production of the High Plains and the Ogallala aquifer. It would be an interesting exercise for an economist to figure out just what the price of food in this country would be if the Ogallala aquifer had remained undeveloped. One thing is for sure—it would be a whole lot higher.

But, as they say, the bill always come due. And in the case of the Ogallala, the bills started coming in around the middle 1970s. As of 1980, water levels in vast areas of the Ogallala in Texas and Oklahoma had declined more than fifty feet. Dire predictions were made about completely depleting the aquifer by the early 2000s. Thoughts of a return to the Dust Bowl days of the 1930s weighed heavily on people's minds, and the wisdom of ever beginning large-scale development of the aquifer was serious questioned.

By the middle 1980s, the Ogallala aquifer was down, but definitely not out. For one thing, it was discovered that the aquifer's natural recharge rates—particularly in the northern High Plains—were a good deal higher than people had first thought. Along the Platte River in Nebraska, for example, water levels have actually risen lately due to surface water irrigation. But the most important thing was that, because of the development of the Ogallala, there was a vast amount of wealth available to develop water-conserving technologies. And these technologies have drastically cut water use in many areas of the High Plains.

For example, the early center pivot irrigation systems operated with very high water pressures, causing water to spray out in a very fine mist—a mist that readily evaporated. In the early days, when water was abundant, this didn't make much difference. Nowadays, to conserve dwindling water resources, center pivot irrigation systems have specially designed nozzles that operate under much lower pressures. The larger water drops formed by these nozzles evaporate at a much lower rate, allowing more than 90 percent of the water to actually reach the roots of the crops being irrigated.

While the mechanical engineers were busy designing improvements into center pivot systems, agricultural scientists were quietly developing improved tillage methods that reduced the need for irrigation water. When the soil is turned over by a plow, it loses most of its stored moisture to evaporation. By reducing or eliminating the amount of tillage, the amount of moisture loss can be drastically reduced.

Because it is the moisture content of the soil—rather than the amount of actual water applied to the soil—that counts in agriculture, another advance has been the development of methods for measuring soil moisture. One of the most elegant is called the "gypsum block." Gypsum (calcium sulfate) is a mineral with a great affinity for water. If it is buried in a soil that is saturated with water, the water will seep into the gypsum, which changes its ability to conduct electricity. By wiring a gypsum block with a couple of electrodes, and burying the block a foot or so in the soil, it is possible to accurately measure the amount of moisture in the soil. In this way, the farmer doesn't have to guess when to irrigate—or not to irrigate—he can know.

But by far the most important technological development have been the farmers themselves. The modern farmer typically runs an operation that is every bit as sophisticated as a multinational corporation. These farmers—at least the ones who manage to remain in business—will use any technology that makes their farms more productive or saves them money. One of the problems caused by dropping water levels is that it takes progressively more and more electricity, and costs more and more money, to lift the water to the surface. When water levels were only 25 feet below ground, it might not have made economic sense to invest in water-saving technology. With water levels 75 or 100 feet below ground—so that it costs double or triple the amount of money to pump the water—investing in water-saving technology becomes an economic proposition. Farmers are not people who willing waste money. As conserving water became more and more economically beneficial, water levels in the Ogallala began to fall less rapidly, and in some cases they are actually rising again. Although many parts of the aquifer system are still at considerable risk, there is now at least a glimmer of hope.

THE STORY OF THE OGALLALA AQUIFER is really a story of American optimism. People knew, in 1960, that large-scale development of the Ogallala would cause depletion of the resource. Yet in the grand tradition of the 1960s, they had faith that once the development was underway and depletion became a problem, some new technology would come along to help solve these new problems. And this is just what happened. There was no way that people in 1960 could have foreseen how improvements in irrigation systems, agricultural practices, and soil moisture monitoring would begin to bring water usage into balance with water availability. These people just assumed that some solution would be forthcoming if it was important enough. They had faith, not only in their own optimistic judgment, but in the ingenuity of their progeny.

Optimism is not the most popular or widely held philosophy in modern America. Nor is the idea that resource *development* can foster resource *conservation* widely held. But the story of the Ogallala aquifer is proof this can occur.

ADDITIONAL READING

Zwingle, E., and J. Richardson. 1993. Ogallala aquifer: Wellspring of the High Plains. *National Geographic* 183(3): 80-109.

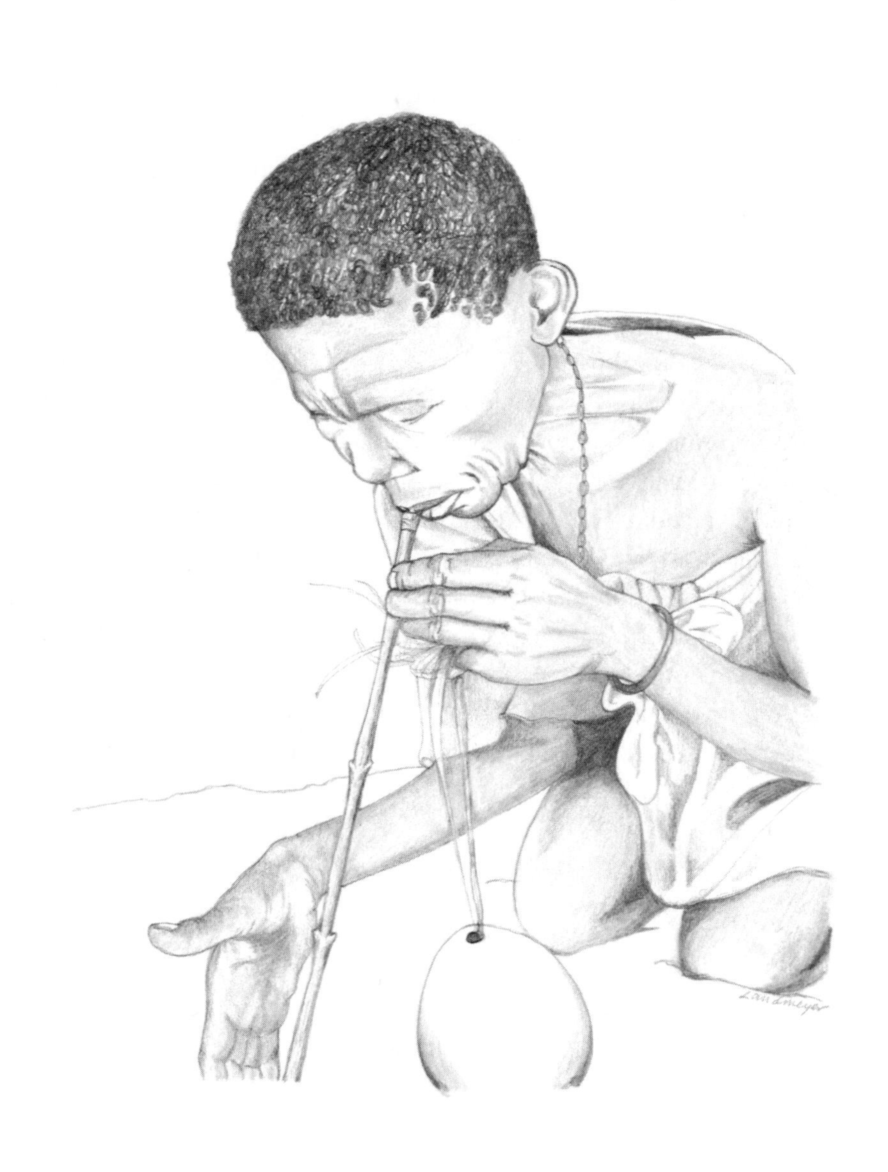

Chapter Sixteen

A Variety of Wells

THE AQUIFERS UNDERLYING the northeastern United States and, say, the Ogallala aquifer of the High Plains don't have much in common. In the Northeast, the aquifer is most often a consolidated granite-gneiss that yields water reluctantly through whatever few cracks and fissures happen to permeate the hard rock. In contrast, the Ogallala aquifer consists of water-saturated sands and gravels that yield water abundantly and enthusiastically. And these are just two of the many types of aquifers found in the United States. Given this variety of aquifers, it is not too surprising that there is as much variety in the kinds of wells that you find from place to place.

This variety of wells is a story in itself.

ACCORDING TO MR. WEBSTER, a well is "a pit or hole sunk into the earth to reach a supply of water." Although this definition may seem somewhat inelegant, it does grasp the two essential features that make a well a well. First, a well is dug, drilled, bored, punched, augered, or scraped into the earth. Second, and more important, the sole purpose of the well is to "reach" and tap ground water. Note that the definition of a well neatly excludes natural springs, seeps, and geysers by the simple fact that they *aren't* "sunk" and that it is absolutely indifferent to the method used to produce the hole. This is a tremendously useful simplification because humans have contrived hundreds of ways for producing holes to act as water wells.

It is a pretty safe guess that the first wells were holes that were simply scraped by hand. Although there is no particular archaeological evidence for this, anthropologists have documented this practice in certain hunter-gatherer cultures. The Kalahari Desert has an extensive network of riverbeds that are dry for most of the year. In places, however, the water table lies within a few feet of land surface. The Bushmen of the Kalahari have learned to find these places by using probes made of hollow reeds. The technique involves poking a reed downward into the sand of the riverbed for two or three feet. The tip of the reed is often stuffed with grass to keep it from being clogged by particles of sand. If water-saturated sand is encountered, the Bushman can use his reed just like a soda straw to suck water out of the ground. For a thirsty hunter who just wants to fill his ostrich eggshell canteen with water, these "reed wells" work just fine. If a family needs a water supply for a few days, the sand may be scraped away to make a shallow and very temporary well.

Such shallow wells have a severe drawback, however. As anyone who has ever dug a hole in the sand at the beach knows, once water-saturated sand is reached, the hole starts to collapse. This makes it impossible to dig any deeper, and severely limits access to the water. Ancient peoples, such as the Mesopotamians and the Chinese, solved this problem by lining the bottom of the hole with stones to keep it from collapsing. The Mesopotamians became so skilled at this they could construct wells penetrating as much as ten feet into saturated sand. This was a tremendous technical achievement. For probably the first time in history, farmers could get enough ground water to irrigate crops. This in turn allowed ancient peoples to grow grain in semiarid wastelands, and was a significant factor in the success of their civilizations.

As impressive as this technical development was, digging wells by hand had several obvious limitations. First of all, the holes had to be large enough to allow the laborers digging them to use a shovel or a pick. This meant that an enormous quantity of dirt had to be dug and lifted to the surface, which was not only difficult work but dangerous as well. If the holes happened to collapse unexpectedly, the diggers could easily be buried alive. The obvious solution was to find a way to dig wells from the surface.

The Chinese were probably the first to use a "churn drill" for reaching ground water. The method consisted of attaching a screw-like bit to wooden rods and twisting the bit into the ground by rotating the rods. The bit would be backed out of the hole every foot or so of its progress, and the "cuttings" removed from the hole with a small scoop. In some cases, the holes had to be cased with hollow logs or bamboo to keep them from collapsing. Needless to say, this sort of drilling was painfully slow. But, working over years and even decades, the Chinese were able to drill wells to depths of hundreds—and in some cases thousands—of feet.

By about 1100 A.D., the Flemish had developed a percussion method of drilling wells. Although just how they learned to do this is not known, the reason the technology developed is perfectly clear. Flanders—what is now Belgium and the Netherlands—was located in the "low country" of Europe, where most of the land is barely above, and in some cases actually below, sea level. In addition, the low country is surrounded by higher land, so that precipitation falling on the surrounding highlands percolates into the ground, recharges the aquifers, and flows naturally toward it. By the time this ground water, trapped underneath relatively impermeable shales and clays, reaches the low country, it is under considerable pressure due to the elevation difference. All that is needed to produce a naturally flowing well is to punch a hole into the underlying aquifer and let the water gush out.

The percussion drilling method used a system of pulleys to raise and drop an iron bit attached to iron rods. The percussive force of the bit ground up the rock and made the hole deeper. Periodically, the bit and rods were removed and the cuttings scooped out of the hole with a specially designed tool. Percussion drilling made it possible to drill wells several hundred feet deep in a relatively short time. The famous wells of Artois, Flanders, (from which we get the word "artesian") were all drilled using percussion methods.

The real technological breakthrough in well-drilling technology, however, was not related to water wells at all, but had to do with oil. In 1859, Edwin Drake drilled the first oil well in Pennsylvania and became wealthy overnight. The fact that people could get rich drilling for oil sparked an explosion of technological development. Percussion methods, which were fine for drilling relatively shallow holes, soon proved ineffective for reaching deeply buried oil. A number of drillers began to experiment with rotary methods. In rotary drilling, a screw-shaped bit was placed on the end of the drilling rods, which were then rotated so that the bit was quite literally screwed into the ground. A gasoline motor (itself a new invention) was used to rotate the iron drilling rods.

Rotary drilling had two large technical problems associated with it. The first was how to remove the "cuttings" as the bit carved the hole downwards. And the second was how to keep the hole from collapsing as it deepened. The solution to both of these problems came in the form of good old American mud. It was discovered that thick mud could be pumped into a hole through the bit and circulated up the hole, past the drilling rods, back to land surface. The weight of the mud served to keep the hole from collapsing. In addition, as the mud circulated up and out of the hole, it would carry the cuttings continuously out of the hole. This method of drilling—called "hydraulic rotary drilling"—was much more efficient than earlier percussion methods. Hydraulic rotary drilling made it possible to tap the deep oil fields of eastern Texas and initiated the great Texas oil rush in the early twentieth century.

It wasn't very long before these new drilling methods were being used to find ground water as well as oil. A good many of the fortune seekers who flocked to Texas during the oil boom ended up broke. But some of these unfortunates also came away with a solid knowledge of hydraulic rotary drilling, and a few of these began to use this new technology for tapping ground water instead of oil. By 1920, for the first time ever, methods were available to easily and routinely probe deep into the earth to find ground water.

BUT PRODUCING A MODERN WELL involves much more than simply drilling a hole into the ground. A well is a device carefully designed to perform three basic functions. The first and most obvious function is to collect ground water efficiently from the geologic formations that the borehole penetrates. A water-saturated formation is just about useless unless that water can be made to drain easily into the well. The second function of a well is, while allowing water to flow in, to exclude the sand and clay particles often present in water-bearing rocks or sediments. After all, a well that produces muddy water is worthless as a source of potable water. And the third function of a well is to prevent surface water from leaking down into the borehole and contaminating the ground water. Rainfall runoff can carry toxic chemicals and bacteria, and is one of the most common causes of contaminated well water.

An enormous amount of technology has developed in the last fifty or sixty years to accomplish these functions. Just how this technology is applied, however, depends largely on the kind of geologic formation being tapped. Ground water-bearing formations are divided first, and most obviously, into *consolidated* formations—those made of hard, solid rock—and *unconsolidated* formations—

those made of loose particles of sand, silt, or gravel. Each of these types presents its own particular opportunities and problems.

Consolidated formations have the enormous advantage that a hole drilled into them will stay open by itself (figure 16.1A). This means that the rock penetrated by the borehole has enough rigidity and strength so that it will not collapse in on itself. The problem with many consolidated formations, however, is that they are not very permeable with ground water flowing mostly through networks of tiny cracks and fractures. Furthermore, these fractures can often be clogged by the process of drilling, which may further reduce the permeability.

The driller can remedy this problem by "developing" the well—by cleaning out the cracks and fissures, and if possible enlarging them, so that water can flow freely into the well. The most common technique for developing hard-rock wells is by "high- velocity jetting." The driller puts a jetting tool into the borehole—usually just a high-pressure hose equipped with a nozzle—and scours the side of the hole with a high-pressure stream of water or air. This blows fine-grained grit and clay out of the fractures so that water can flow more freely into the well. After jetting, the well is usually pumped vigorously to draw water into the well. By alternately jetting and pumping, the yield of an open-hole well can often be doubled or even tripled.

In unconsolidated formations, techniques for building wells are more involved (figure 16.1B). After the borehole is drilled, and the hole is being held open by thick drilling mud, the first order of business is for the driller to "screen and case" the well to keep it from collapsing. While the mud is still in the hole, the driller lowers a well screen into position next to the most permeable sands or gravels encountered by the borehole. As the name implies, the "well screen" consists of wire screen wrapped around metal rods, which allows water to flow but keeps out larger sand particles. The well screen is attached to a string of "casing" designed to keep the upper part of the well from collapsing.

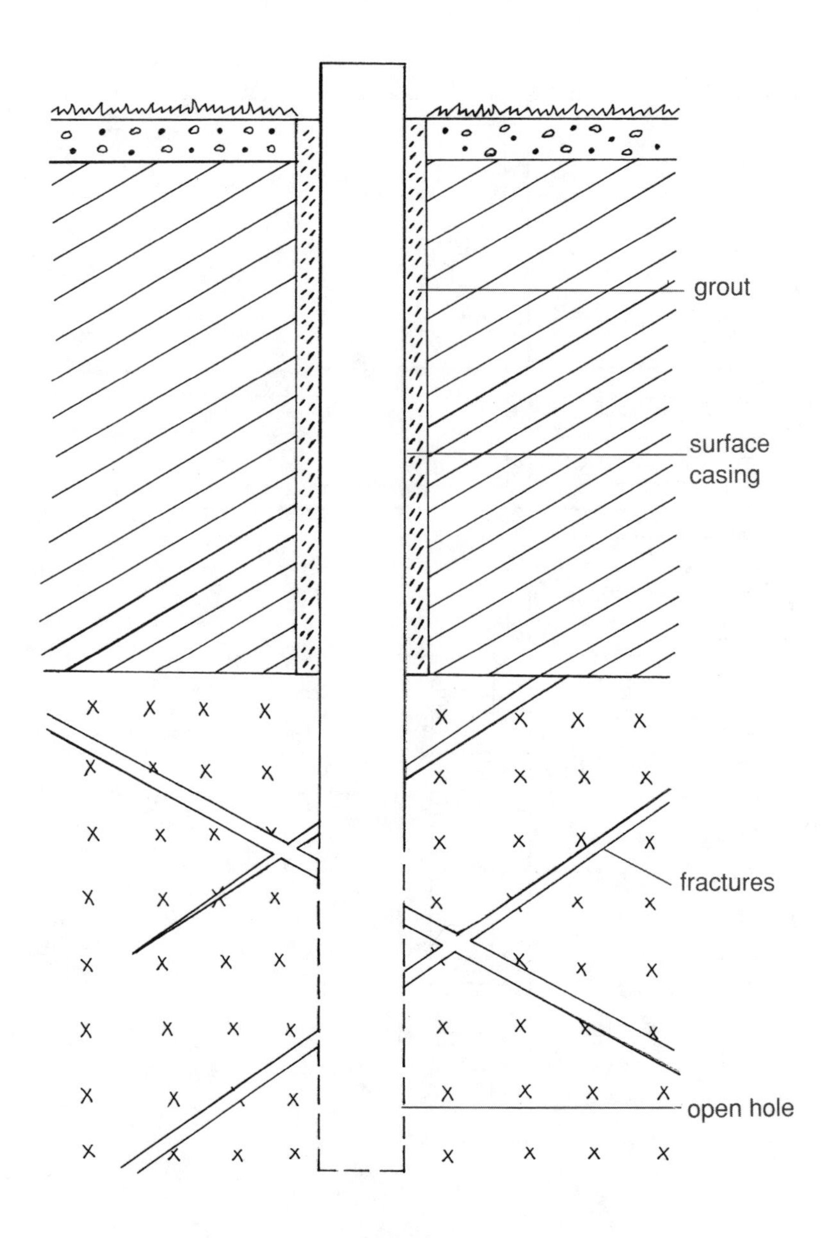

Figure 16.1A
Cross section of a typical well in a consolidated aquifer.

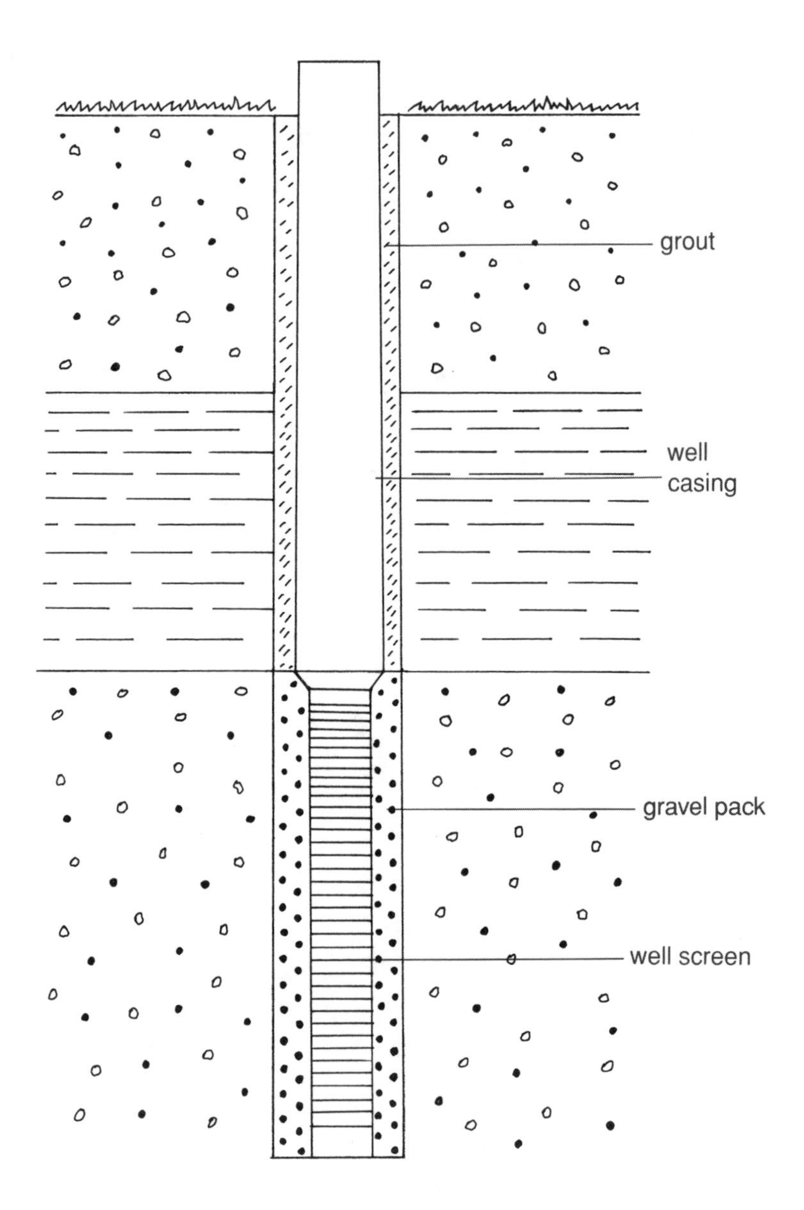

Figure 16.1B
Cross section of a typical well in an unconsolidated aquifer.

Once the screen is in place, the driller packs the space between the screen and the formation with gravel carefully sieved to a uniform size. This "gravel pack" is designed to help water flow easily into the well, but just as important, it helps filter out small clay and silt particles that might be present in the formation and which must be kept out of the water flowing into the well.

Once the screen, casing, and gravel pack are in place, the well is ready for development, generally a more involved procedure for unconsolidated than consolidated formations. One reason is that, as with consolidated formations, the most common method used to remove traces of drilling mud in the screen and gravel pack is high-velocity jetting, followed by pumping. Applied to unconsolidated sediments, this is often a tedious operation that can take one or more full days, but it is essential if the completed well is to be efficient and free of suspended sand or clay. For high-capacity municipal or industrial wells, it is not uncommon for drillers to spend a week or more developing a well in an unconsolidated formation.

The final step in completing a well in either consolidated or unconsolidated geological formations is to seal it off so that storm water at land surface cannot run down the casing and contaminate the well. This is accomplished by "grouting" the well. In a consolidated formation where casing and screens aren't necessary, a section of ten or twenty feet of surface casing is placed into the borehole to be grouted. During grouting, the driller fills the space between the well casing and the borehole with a grout made of impermeable bentonite clay, cement, or both. In most cases, the driller uses small-diameter pipe to place the grout firmly around the casing, and the well is grouted only to the bottom of the casing. In an unconsolidated formation, by contrast, it is good practice to grout the well from the top of the well screen all the way to land surface, which prevents bacterial or chemical contaminants from leaking into the well. The variety of wells—from open-hole wells in fractured granite, sandstone, or limestone to screened wells in the marine sediments of the Atlantic Coastal Plain or the alluvial sediments of the San Joaquin Valley—is a reflection of the geologic diversity of the United States. It also reflects the full spectrum of human uses for ground water. Wells produce water—in varying quantities—for dozens of reasons. They produce water for animals and crops, they provide process water for chemical plants and cooling water for power plants, they drain shallow sediments so that building foundations can be poured, they ring hazardous waste sites in order to monitor the migration of contaminants, they keep mines from being flooded, they are used together with heat pumps to heat and cool buildings. And, most important, they provide drinking water in varying quantities for individual homes, subdivisions, towns, and cities throughout the country.

It is quite fitting that Mr. Webster settled on such a simple definition of what makes a well a well. Otherwise, it is just too complicated.

ADDITIONAL READING

Johnson Division. 1986. *Groundwater and wells.* 2d ed. Edited by Fletcher G. Driscoll. St. Paul: Johnson Division. 1,089 pp.

Van der Post, L., and J. Taylor. 1984. *Testament to the Bushmen.* New York: Viking Penguin. 176 pp.

PART THREE

Evil and the Wells

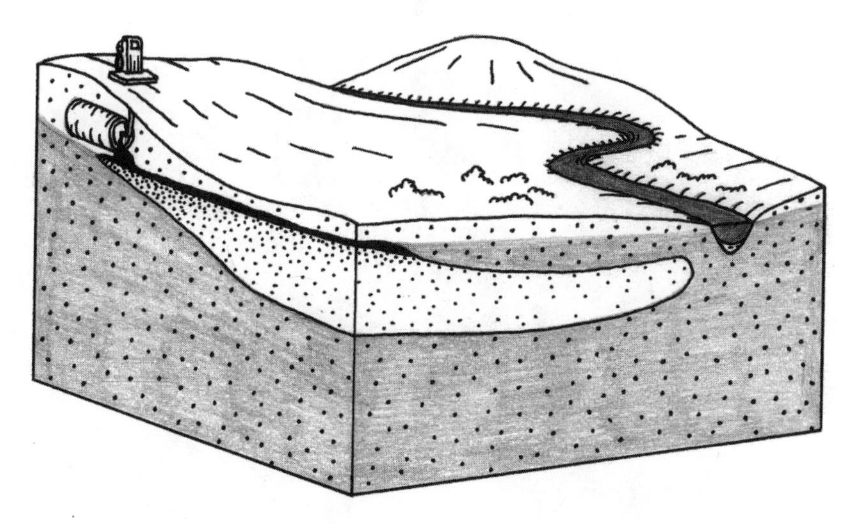

But as habitations were gradually built up and the population increased, it was noticed that the water in the wells, especially in the more populous portions, was rapidly losing its pristine purity, and was becoming hard, impotable, and injurious to health. . . . As the evil continued to increase in the wells, rain-collecting cisterns came into general use. . . .

Municipal Report of the City of Charleston, South Carolina (1881)

Chapter Seventeen

From Pristine to Poison

THE CONTAMINATION OF GROUND WATER by human activities is not, as some people might think, a particularly recent phenomenon. Very early in the history of ancient Rome, for example, it was noticed that shallow wells commonly used for water supply often became hopelessly contaminated. Much of this contamination came from the open-hole latrines and privies that disposed of sewage wastes in those days. The Romans soon concluded that ground water was an unreliable source of water for human consumption. They accordingly spent huge amounts of money building public aqueducts to bring in clean, pristine surface water from the nearby mountains. The Romans realized, as have every civilized people since, that living in cities is impossible if the water supply is not reliably clean and fresh.

The experience of the Romans has been repeated over and over in history. When people use shallow ground water systems both as sources of drinking water and as repositories for human waste, it is just inevitable that ground water will become contaminated. This was true long before Superfund legislation (1980) was enacted to clean up ground water contamination, and doubtless will be true in the future as well.

The experience of Charleston, South Carolina, is as good an illustration of this general principle as any.

THE FIRST EUROPEAN COLONISTS OF SOUTH CAROLINA did not have an easy time. Upon arriving, they had built a settlement on the west bank of the Ashley River, but from the beginning, that turned out to be a poor choice. The site was low-lying and damp, and the colonists suffered disproportionally from a variety of fevers and sickness. What was worse, it had no natural defenses, and the occasional incursions of the native population weighed heavily on the minds of the settlers. So, in 1678, the decision was made to abandon the original site and build a new settlement, to be named Charles Town, on a peninsula between the Ashley and Cooper Rivers.

The new site soon proved to be a vast improvement. For one thing, Charles Town, or Charleston as it came to be called, had more elevation and drainage was better. And, being on a peninsula, it was much easier to defend. But more important, Charleston's peninsula was underlain by a bed of quartz sand 18 to 20 feet thick. Below this sand was a thick bed of impermeable blue-black clay (figure 17.1). The clay acted as a natural water trap for rainfall, and the quartz sand was so permeable that it readily yielded water to shallow wells. The new residents of Charleston could dig a well twelve or fifteen feet deep just about anywhere and have an unfailing supply of fresh, clean water.

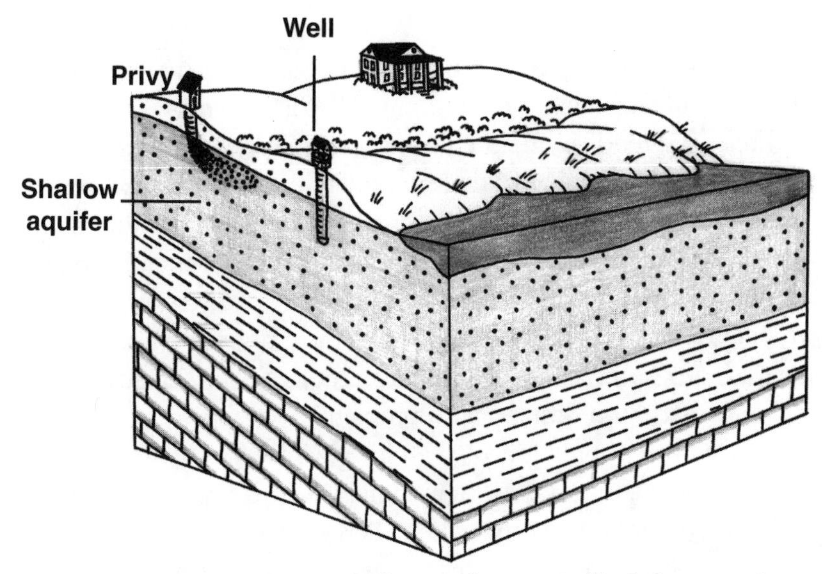

Figure 17.1 How Charleston's shallow squifer was used both for a repository for sewage wastes and as a source of fresh ground water in the 1700s.

These happy circumstances prevailed for fifty years or so, during which time the little town grew steadily. By the middle of the eighteenth century, however, the character of the water available from the shallow wells began to change noticeably. Wells began producing water that was rust-colored, foul-smelling, and unfit to drink. The residents of Charleston referred to this as an "evil" afflicting their wells. What was really happening was that untreated human sewage was seeping out of hundreds of unlined outhouses and privies in the city. These wastes were migrating down to the water table and contaminating the ground water.

It is worth noting that this practice didn't cause any particular problems for the first fifty years of Charleston's history. The reason for this is that all ground water systems have a certain capacity to cleanse and renew contaminated water. Sewage effluent seeping from privies contains dissolved and particulate organic carbon, nitrogen compounds, and a variety of bacteria and viruses. As this effluent percolates through fine quartz sand, particulate carbon and microorganisms are filtered out. The dissolved organic carbon and nitrogen tend to migrate farther but are progressively metabolized by microorganisms that naturally inhabit shallow ground water systems. This converts potentially toxic compounds into harmless carbon dioxide and nitrogen gas. In general, shallow aquifers will return small quantities of untreated sewage to clean, pristine water fairly quickly. As long as the amount of sewage effluent did not exceed the "assimilative capacity" of the underlying aquifer, Charleston didn't have a problem.

But as the population of the city grew, and as the amount of sewage effluent increased, the assimilative capacity of Charleston's shallow aquifer was overwhelmed. Rather than being filtered out, pathogenic bacteria and viruses present in sewage were transported to nearby wells. And rather than being metabolized, organic carbon and nitrogen compounds accumulated in ground water and produced a distinctive foul smell. Iron oxides present in the sands dissolved and leached into the contaminated water, turning it a gentle—if unhealthy—shade of red. By 1800, most of the wells in Charleston had been abandoned, and rain-fed cisterns became the most common water supply for the Charlestonians.

IT IS NO SECRET THAT PERCEPTIONS do not always reflect reality. In the case of Charleston's poisoned wells, the original perception was that the shallow aquifer could act both as a source of drinking water and as a repository for human sewage indefinitely. The reality was that once the assimilative capacity of the aquifer was exceeded, the ground water would inevitably become contaminated. That reality never changed in the years between the founding of the city and the eventual abandonment of the poisoned wells. What did change was the perception of that reality. The Charlestonians were some of the first Americans to experience what has come to be called "environmental realism." Environmental realism is simply the process of adjusting changeable, and frequently erroneous, perceptions to match the inflexible realities of the earth.

The experience of Charleston in the 1700s mirrors more recent American history. From the day Native Americans first set foot in North America 12,000 years ago, human wastes have been delivered to shallow ground water systems, but populations were so small that the wastes never came close to overwhelming the available assimilative capacity of these aquifers. Even when Europeans began arriving in force, it was rare for populations to be so large (like Charleston's) as to exceed this assimilative capacity.

In the twentieth century, however, two important changes occurred. First, the population grew so rapidly that the volume of wastes being generated exploded and in many places began to exceed the available assimilative capacity of the underlying aquifers. Second, we began producing an extraordinary variety of new kinds of chemical wastes. Where once there had simply been sewage effluents, now there were toxic metal-bearing effluents from mines, solvents from machine shops, fuels and oils from the newly invented automobiles, pesticides, herbicides, chemical manufacturing wastes, sludges, coal tars, and transformer oils. Given this assault, it is little wonder that by the 1970s the assimilative capacity of many ground water systems had been overwhelmed.

And, as was the case in Charleston, perceptions lagged considerably behind reality. At first, dumping chemical wastes into unlined settling basins, or actually pumping them directly into wells, was not perceived as a problem. But gradually, as instances of people being poisoned by drinking contaminated well water became more and more common, perceptions began to change. The environmental reality was that these sorts of waste disposal practices could kill people, and that these practices had to stop.

But as grim as the situation seemed in 1976, when Congress enacted the Resource Conservation and Recovery Act (RCRA) to stop shoddy waste disposal procedures, and in 1980, when Congress enacted Superfund legislation to begin cleaning up accumulated contamination, it was not and is not a hopeless situation. All ground water systems have a capacity to assimilate and purify contaminated water. This capacity, which varies widely between different aquifers and between different kinds of contaminants, often makes it possible to clean up contaminated ground water. That environmental reality, however, is often as hidden as the ground water systems themselves.

Take, for example, Weldon Spring, Missouri.

DANGER
TNT-
CONTAMINATED
SOIL

Chapter Eighteen

Inattention, Ignorance, or Ill Will

In early 1942, the national survival of United States was very much at risk. The Japanese were sweeping across the Pacific Ocean toward Australia and Hawaii, and Hitler's armored divisions were advancing steadily across eastern Europe. Millions of people were dying around the world, and many Americans fully expected a Japanese invasion of California. In these grim days, the economy of the United States shifted immediately to a war footing. War production, particularly the production of weapons and ammunition, became the nation's top priority.

Weldon Spring, Missouri—located about 40 miles west of St. Louis—had once been a sleepy farming community nestled between the Missouri and Mississippi Rivers. But with the clouds of war looming dark on the horizon, it underwent a dramatic transformation. In 1940 and 1941, the Department of the Army procured about 17,000 acres of land around Weldon Spring, and proceeded to build the largest TNT-producing facility in the world. In the process of acquiring this land, three whole towns—Howell, Toonerville, and Hamburg—disappeared from the face of the earth. By 1943, when production was at its peak, the Weldon Spring Ordnance Works employed 5,000 workers and was producing a million pounds of TNT each and every day.

During this time, the facility was operated under the certain knowledge that the United States was fighting for its life, and that the production of explosives was a critical factor in that fight. Because of this, the plant managers were concerned exclusively with producing the TNT that the War Department needed and loudly demanded. All other issues were subordinate to TNT production. Workers at the site were routinely exposed to significant dangers—principally explosion hazards and caustic chemicals—and three workers died in accidents during the war.

Given these pressing problems, it is not too surprising that the accumulation and disposal of chemical wastes was not a top-priority issue. The original design for the facility called for the construction of settling tanks and treatment plants to deal with chemical wastes. Unfortunately, construction of these tanks and plants was behind schedule when TNT production began in November of 1941. Because of the urgent need for TNT, the plant operators dug holding ponds to temporarily hold the wastes. Even after the treatment plants were finished, however, settling tanks often overflowed, wastewater lines clogged, and pumps failed. Nobody was particularly happy about this state of affairs, but nobody was willing slow down TNT production in order to fix the problems either.

Trinitrotoluene, TNT for short, is manufactured from just two chemicals: toluene (figure 18.1A), more commonly known as an important component of gasoline, and nitric acid (HNO_3). The process for making TNT involves sequential treatment of toluene

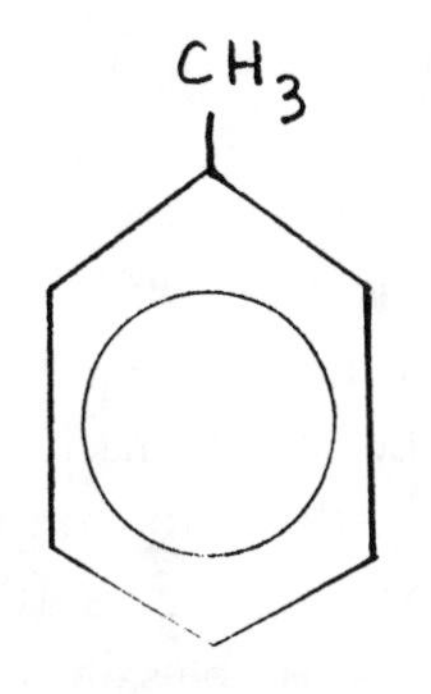

(A) The chemical structure of toluene.

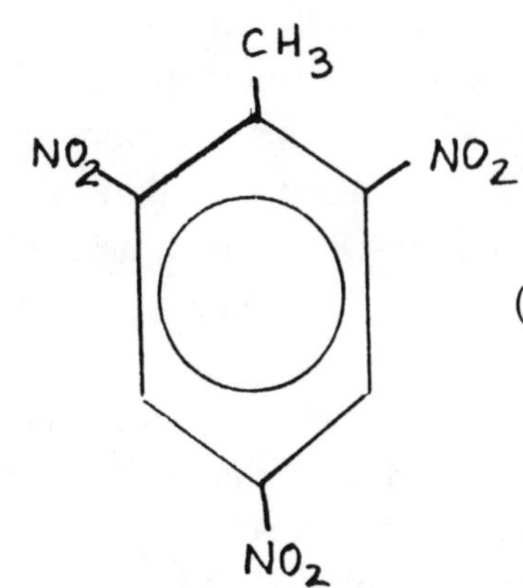

(B) The chemical structure of 2,4,6-trinitrotoluene.

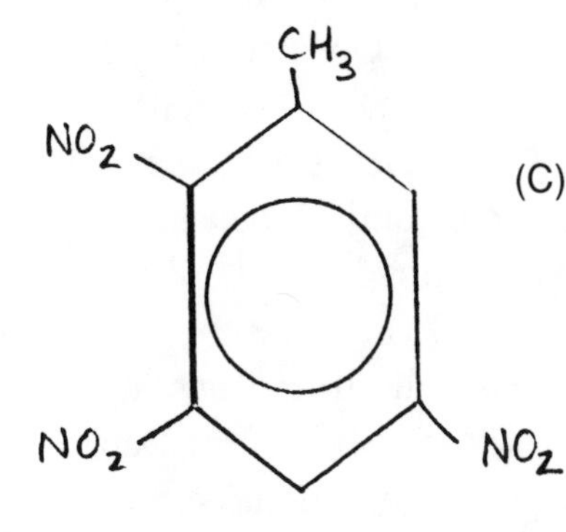

(C) The chemical structure of 3,5,6-trinitrotoluene, a chemical impurity formed in the manufacturing process.

Figure 18.1

Chemical structures of toluene and several different TNT isomers.

with nitric acid. This treatment serves to attach "nitro" (NO_2) groups to the toluene molecule, producing the TNT molecule (figure 18.1B). The three nitro groups, which are located in the 2, 4, and 6 positions (counting clockwise around the molecule), are the most characteristic feature of the molecule. Hence its common name "trinitrotoluene."

Chemical synthesis, however, is an inexact business, and there is little reason why the nitro groups should all behave themselves and fit neatly into the 2, 4, and 6 positions. Some of the nitro groups, for example, might slip into the 3, 5, and 6 positions (figure 18.1C). These "isomers" of TNT, so called because they have the same chemical formula but a slightly different structure, do not ignite with the same efficiency as the 2,4,6-molecule and are considered impurities.

Because of these impurities, an important part of the manufacturing process was to purify the desired TNT. This was done by "washing" the final products with a solution of sodium sulfite. The desired 2,4,6-isomer stayed behind as a crystalline solid, and the impurities—referred to collectively as "nitroaromatic" compounds—were washed away as a waste called "red water."

Over the years, uncounted millions of gallons of waste red water were generated at the Weldon Spring facility. Much of this waste water simply escaped from the leaky wastewater treatment facilities and ran off into gullies and ditches, where it percolated into the ground. In addition, red water leaked steadily onto the soils underlying the "wash houses," where the TNT purification took place. Over the years, TNT and other contaminants accumulated, until some of the soils contained as much as 10 or 15 percent TNT. Not surprisingly, these soils acquired the color of the red water, and some of them remain red to this day, fifty years after the contamination occurred.

IT IS WORTH WONDERING, AT THIS POINT, just who the villains in this story were. It is pretty hard to pin the "villain" label on the Department of the Army. After all, they were simply trying to defend the nation in a time of dire national emergency. Were the plant managers and workers—who, incidentally, were civilian contractors—the villains? It was their waste disposal practices that resulted in contaminating 1,700 acres of land with huge amounts of potential carcinogens. But these managers and workers were doing a tough job under extremely hazardous conditions. They worked in lead-lined buildings and wore nonsparking boots and clothing so that the ubiquitous TNT dust would have less chance of exploding beneath their feet. When you consider that these people (many of them women) were willing to work under such dangerous conditions to produce desperately needed ordnance, it is hard to label them as "villains" either.

The answer is simply that there were no villains. Like the Charlestonians three centuries before, the people at Weldon Spring produced a severe environmental problem out of inattention and ignorance rather than ill will. They didn't have the time or the leisure to consider the environmental consequences of what they were doing. But even if they had, it is doubtful that much would have been done differently. The ability to mass-produce chemicals like TNT was a technology that was fairly new in 1940. The people of the time had no experience with the residual effects of TNT contamination. They had no idea that red water leaking into the underlying ground water system would carry nitroaromatics miles from the site, contaminating both wells and springs. Furthermore, they had no idea that chronic exposure to nitroaromatics could lead to liver damage or cancer in humans.

They just didn't know.

But we do know. We know because of our collective experience from Charleston in the seventeenth century to Weldon Spring in the twentieth century. It is entirely true that unscrupulous, greedy individuals create environmental problems for no other reason than personal or corporate gain. But the lesson of history is that many environmental problems—and particularly many ground water contamination problems—are caused more by ignorance and inattention. This doesn't make these problems any less serious, or the contaminants any less toxic or easier to clean up. But it does indicate how such contamination can be prevented in the future.

The evil, it seems, is in the ignorance.

Chapter Nineteen

The Riddles of Mother Earth

The fact that the TNT contamination at Weldon Spring and many other instances of environmental contamination were not perpetrated by rascals or villains does not make them any less serious. Just because the TNT-laced soils—and the ground water contamination coming from these soils—were produced by well-meaning, hardworking, patriotic souls does not make them any less carcinogenic or any less of a danger to the community. Regardless of how the contamination got there, it needs to be dealt with and, where possible, cleaned up.

Cleaning up environmental contamination—particularly the product of what would now be considered incredible negligence—is always a tedious and usually a frustrating endeavor. Environmental laws are labyrinthine in their complexity, and sorting out various liabilities can take years of legal wrangling. But, curiously enough, it is not just the twists and turns of legal reasoning—and the resulting tangled web of liability—that get in the way of cleaning up contaminated sites. Often, the biggest problem is simply understanding the hidden quirks, oddities, peculiarities, and eccentricities of the underlying ground water systems.

And Weldon Spring is a classic example. By the end of World War II, the Weldon Spring site was a mess. Years of pell-mell TNT production, using relatively crude technology, had left large areas of the site literally coated with TNT and other nitroaromatics. Eyewitness accounts of the site after it was taken out of production describe the heavily contaminated soil underneath the "wash houses" where the TNT was purified. They also describe places on the site where the overflow of process wastewater had produced baseball-sized chunks of crystalline TNT. This TNT was not pure enough for making ordnance, and so it was simply spread out on the ground or dumped into trenches.

When the site was decommissioned in 1945, the Army decided to clean it up, using the best technology available at the time. This "technology" was simply to burn down the buildings that were heavily contaminated with TNT, and burn whatever chunks of crystalline TNT they could find. Because no one knew what to do with the red water-contaminated soils, they just left them where they were. And because no one was aware of the contaminated ground water underlying the site, that wasn't even an issue. The Army then packed up whatever equipment was salvageable, closed the site up, and just let it sit.

And the site just sat for the next twenty years or so. During that time, those parts of the original reservation that had not been used for TNT production were returned to state and local control. Much of this uncontaminated land was converted into wildlife areas, and another part was converted into a training facility for the Army and for National Guard Units. Control of about 1,700 acres of TNT-contaminated property—along with the responsibility to clean it up—was turned over to the Army Corps of Engineers in 1969.

The year 1980 was a pivotal one for the Weldon Spring site: Congress passed the Comprehensive Environmental Response, Compensation, and Liability Act (CERCLA), known universally as "Superfund." This legislation set out procedures for deciding who was legally responsible for environmental contamination, and procedures for initiating and completing site cleanups. It committed a billion dollars in funding to get cleanups started at high-risk sites. Soon, Weldon Spring was placed on the EPA's National Priorities List as one of these high-risk sites, and the Army Corps of Engineers was told to initiate site cleanup without delay.

But it is one thing for lawyers in Congress to mandate cleaning up contaminated sites, and it is quite another to actually do it. For one thing, nobody was really sure where the contamination was. In the thirty years since the site closed, most of the people familiar with site operations had left or died, and a thick forest had grown up, effectively hiding much of the contamination. The first order of business, therefore, was to do a comprehensive assessment of the contamination that was present.

This was an enormous undertaking. In the next few years, tens of thousands of soil samples were taken and analyzed for TNT and other nitroaromatic compounds. The extent of the soil contamination was painstakingly mapped, the depth of contamination determined, and the volume of contaminated soil estimated. In addition, monitoring wells were drilled, ground water samples analyzed, and the extent of ground water contamination assessed.

However laborious, finding the soil contamination was fairly straightforward, but it was soon apparent that something odd was going on with the ground water. Low concentrations of nitroaromatic compounds, on the order of one or two parts per billion, were turning up in wells and springs miles away from the site, which seemed to suggest that there was an enormous plume of heavily contaminated ground water underlying the site itself. The odd thing was that concentrations of nitroaromatics found in ground water beneath the site were often just as low. If the site were the source of contaminants reaching the distant wells and springs, you would expect the concentrations to be much higher. Did this mean that there could be some other source of contamination?

Another odd thing was the kind of contamination found in the ground water. Something like 99 percent of the soil contamination was TNT. If the soil was the source of contaminants in ground water, you would naturally expect TNT to be the most common contaminant. But this was not the case. The most common contaminants found in the ground water were dinitrotoluene (DNT), which has only two nitro (NO_2) groups on it (figure 19.1A), or aminonitrotoluene (ANT) compounds ("nitroamines"), which have amino (NH_2) groups as well as nitro groups attached to them (figure 19.1B). Again, did this mean that there was some other unknown source of contamination? By 1990, these and other questions had the Corps of Engineers totally baffled, which in turn made it very difficult to proceed with the cleanup operations.

At this point, however, the corps got a bit of luck. They managed to secure the help of a particularly sharp hydrogeologist named John Schumacher who was working for the U.S. Geological Survey, and he went right to work trying to figure out what was going on.

The first thing Schumacher did was take a hard look at the kind of ground water system he was dealing with. The Weldon Spring site is underlain by massive limestones known locally as the Burlington and Keokuk formations (figure 19.2). Because these limestones are tightly cemented, they have very little primary porosity. But they are extensively fractured, and ground water moves easily along these fracture traces. What is more, as ground water moves along these fractures, it gradually dissolves the limestone, creating large fissures, so large in places they act literally like underground rivers. In most ground water systems, water seeps along at a leisurely pace of a few feet per year. At Weldon Spring, however, ground water can move hundreds of feet in a single day.

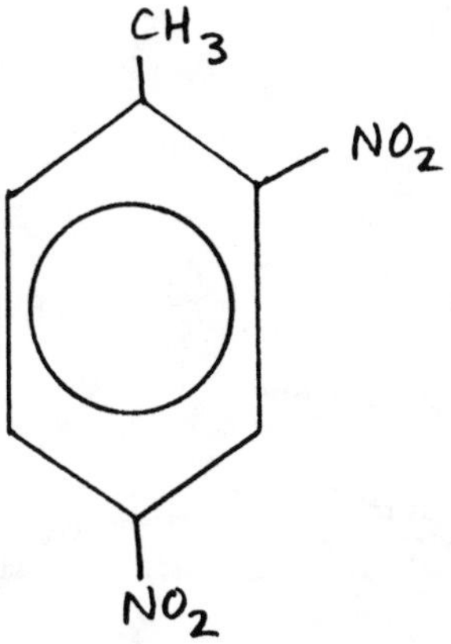

(A) 2-4 dinitrotoluene.

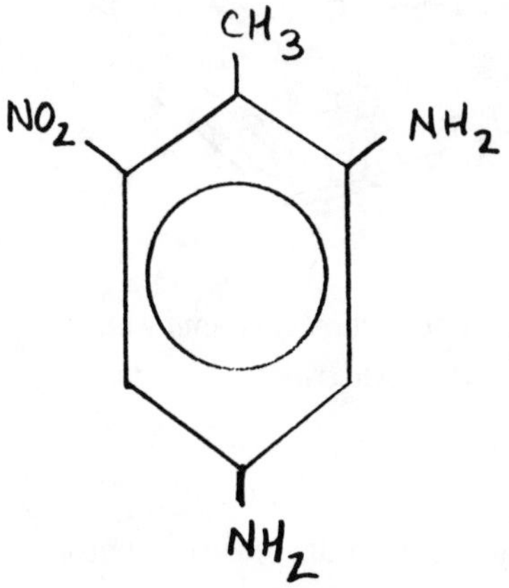

(B) 2-4 aminonitrotoluene.

Figure 19.1 Chemical structures of a dinitrotoleuene (DNT) and an aminonitrotoluene (ANT) compound.

This immediately suggested an answer to the puzzle of why contaminants were found so far from the site. The fissures were literally and quickly "piping" the contaminants away from the site. The good news was that there wasn't necessarily another unknown source of contamination. The bad news was that the contaminants

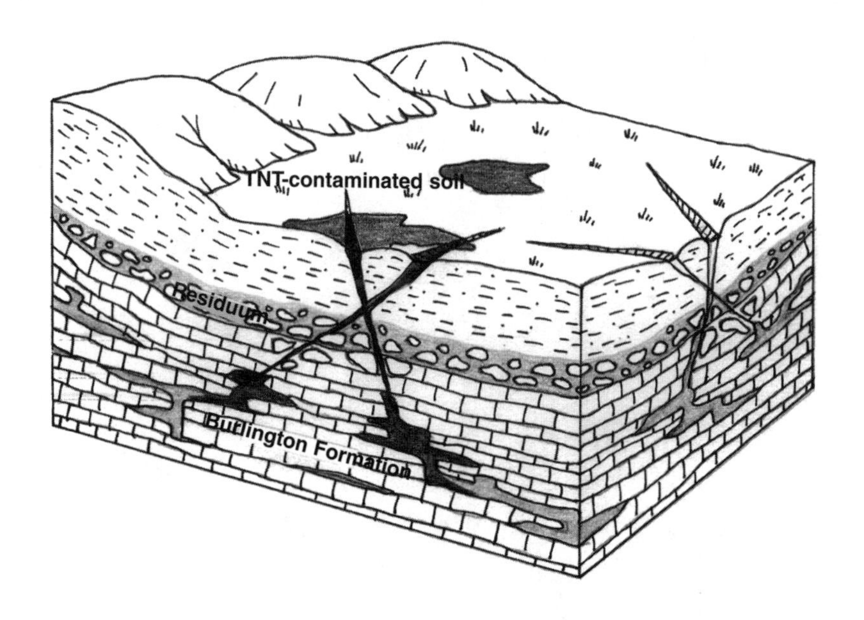

Figure 19.2 Nitroaromatic contamination of ground water at Weldon Spring, Missouri.

were being spread quickly, suggesting that ground water was a more likely exposure pathway for humans than had previously been thought.

Where the DNT and the nitroamines were coming from was a tougher puzzle to solve. If the contaminants in the ground water were coming from TNT-contaminated soils, as Schumacher believed, then something must be happening to the nitroaromatics between the time they leached from the soil and the time they finally entered the water table. In order to find out just what was happening, Schumacher placed a series of soil moisture collectors at different depths below one particularly contaminated area. He then collected soil water samples from different depths and analyzed them for nitroaromatics. At a depth of one or two feet, TNT was just about the only dissolved contaminant present. But at depths of three, five, and ten feet, TNT concentrations decreased and concentrations of the nitroamines and DNT increased. Apparently, the TNT being leached out of the soil was being transformed by some process. The problem was, Schumacher had no idea of what process that might be. He had effectively replaced one mystery with another.

But John Schumacher was not one to give up so easily. He knew from reading the literature that soil microorganisms could transform TNT to other compounds. Could this be happening at Weldon Spring? And even more tantalizing, could these transformations be going on in the limestone aquifer as well?

To find out, Schumacher recruited the help of a particularly able microbiologist named Paul Bradley, and supplied him with samples of the soils and the limestone from a contaminated site. In a series of laboratory experiments, Bradley dosed the soil and rock samples—and thus the microbes present in them—with TNT. He also sterilized a set of samples, dosed them with TNT, and then systematically measured how the TNT changed over time. In the sterile samples, nothing at all happened, and the TNT remained untransformed. In the "live" samples, however, TNT was quickly transformed to a series of nitroamine products, and then disappeared altogether. Furthermore, the progression of TNT to nitroamines was virtually identical to what Schumacher had seen in his soil water samples.

But Bradley found something else out as well. If he dosed the live samples with particularly high concentrations of TNT—anything over about 100 parts per million—the TNT was not transformed at all. When he looked closer, he found out why. The high concentrations of TNT—which, after all, is toxic—were killing the microorganisms. That explained why the soil TNT, present in concentrations up to 10 percent (100,000 parts per million), was not being degraded: it was killing the soil microorganisms. But if small amounts of TNT were leached into the soil water and percolated downward, the concentrations became low enough for microorganisms to tolerate, which in turn allowed them to effectively degrade the TNT and nitroamine contaminants.

Now, all of the sudden—and thanks largely to John Schumacher and Paul Bradley—there was a perfectly obvious strategy for cleaning up the ground water at Weldon Spring. The trick would simply be to cut off the source of contaminants leaching into the underlying aquifer. If this was done, the contaminants already in the ground water—and present at such low concentrations that they were no longer toxic to microorganisms—would quickly be degraded. In the absence of a TNT source, the aquifer would be returned to pristine conditions within a few years.

Early in 1995, the Army Corps of Engineers, the EPA, and the State of Missouri agreed to a plan for cleaning up the Weldon Spring site. In this plan, the heavily contaminated soils would be excavated and decontaminated by either incineration or some other appropriate process. Removing the contaminant source would allow the natural microorganisms present in the aquifer to renew the ground water like a giant septic field. The progress of this natural renewal would be closely monitored over the years to verify that it was actually working. There were problems, of course. The corps favored incineration as the treatment process for the contaminated soils. This didn't exactly thrill the local residents, who were worried about air pollution. In addition, there were the usual tiresome arguments about permits and timetables and interim reports and dozens of other details. But overall, everybody involved agreed on the strategy.

THE STRANGE AND OFTEN INEXPLICABLE BEHAVIOR of ground water has alternately fascinated and frustrated humankind throughout much of history. But the story of Weldon Spring shows that these oddities can be well worth puzzling through. By patiently working out the pathways that delivered nitroaromatic compounds to the underlying limestone aquifer, and by doggedly investigating how microorganisms transformed these contaminants, a workable and sensible cleanup strategy had been identified.

The riddles of Mother Earth, it seems, are also opportunities.

Chapter Twenty

Claystone, Kitty Litter, and Hazardous Wastes

WELDON SPRING IS AN ALL-TOO-COMMON EXAMPLE of what happens when hazardous wastes are not handled and disposed of properly. It is reasonable to ask, therefore, just what constitutes "proper" disposal of such wastes. Prior to the 1970s, the answer to that question would have been much different from what it would be today. This is because much more has been learned and much more technology is now available. Like any technology, however, it had to be learned step by painful step.

One example of how this learning process played out happened near a sleepy southern town named Pinewood, South Carolina.

THE PINEWOOD SECURE LANDFILL has the distinction of being one of the largest operations in the eastern United States for handling and disposing of hazardous wastes. Exactly what constitutes a "hazardous waste" is actually pretty involved. Suffice it to say that a variety of chemical residues from industrial processes—residues that are either too noxious or too expensive to recycle—make their way to Pinewood on a regular basis. The landfill provides a needed service because, without a place to dispose of these wastes, many industrial operations across the country would be forced out of business.

The real story of the Pinewood Secure Landfill, however, does not begin with hazardous wastes, but with something altogether more mundane. It begins with kitty litter, the absorbent stuff used by cat owners in litter boxes.

In the early 1970s, an ambitious and enterprising geologist named Paul Bennett noticed the presence of some unusual rocks near Pinewood, South Carolina. Just how he got interested in these rocks is a mystery to this day, but it may be that he first saw them exposed in a road cut. The rocks were composed of very fine-grained clays that, to a casual observer, seemed ordinary enough. Bennett, however, noticed that these rocks were much less dense than ordinary claystones. The rocks were so light, in fact, that they would almost float in water. Intrigued, Bennett took a closer look and discovered that it wasn't just a garden variety claystone at all, but that the clay particles were cemented together by opal—the gem name for a mixture of silica and water ($SiO_2 \cdot 4H_2O$). Because of this, the rock is still known today as an "opaline claystone."

The actual origin of the opaline claystone, especially the origin of its unusually low density, is not well understood. The claystone was deposited during the Eocene epoch, about 50 million years ago. The fact that its clays are so fine-grained suggests they settled out of a body of water that was largely stagnant, perhaps a saline bay cut off from the sea. And because it was stagnant, the water also appears to have been excessively acidic or basic, either of which condition would tend to leach silica (SiO_2) from the clays. As a result of leaching, the rock lost much of its original mass (about 25 percent), which explains its low density and which accounts for the presence of the opal (some of the dissolved silica remained behind as an opaline cement).

But what really makes the opaline claystone valuable is that it has such a high capacity for absorbing liquids. As the silica was leached from the claystone, the claystone developed an unusually high porosity—that is, the spaces between clay particles became unusually large—and therefore also the ability to trap large amounts of liquid. Opaline claystone from Pinewood, once it has been dried, crushed, and sieved, can absorb up to twice its own weight in water or any other liquid. Because of these absorbent properties, Bennett decided that the claystone could be used to manufacture an excellent kitty litter.

And so, in 1972, Bennett leased 100 acres of land near Pinewood that was underlain by the claystone and went into business quarrying, drying, crushing, and sieving opaline claystone and selling the product as kitty litter. Because the claystone was so near the surface, quarrying it was relatively easy and could be done with an ordinary backhoe. The most expensive part of the operation was drying the claystone, which required fuel to fire the drying kilns. Unfortunately for Bennett, the Arab Oil Embargo began just as he was getting his operation going, and fuel oil suddenly became extremely expensive. This, in turn, threatened to put Bennett out of business.

But Bennett was a resourceful fellow, and was not going to give up easily. Because he didn't need a particularly high grade of fuel to dry the claystone, he began to use waste oils from nearby manufacturing plants to fire his kilns. And because these waste oils could be had just for the cost of hauling them off, it was extremely economical, and Bennett's kitty litter business began to make money. For a while, several name brands of kitty litter were being packaged at Bennett's operation.

It was at this point that Bennett's business began to undergo a curious metamorphosis. The waste oil–producing industries did not produce just waste oil, but other toxic liquid wastes as well. Furthermore, these industries were just as anxious to be rid of these other toxic wastes as they were to be rid of the waste oil. Apparently, one of Bennett's suppliers told him that if he wanted the waste oil for free, he would have to haul off some of these other toxic wastes—things like used degreasing agents, metal-plating residues, and paint thinners—as well.

Bennett considered this new development. On the one hand, he needed the free waste oil to stay in business. On the other, what would he do with barrels of things like used degreasing agents? As it happened, Bennett had another vexing problem. Not all of the processed claystone he produced was of high enough quality to sell as kitty litter, and this inferior dried claystone was just accumulating into piles. Perhaps, Bennett thought, this excess claystone could be used to solidify the liquid toxic wastes and then used as landfill in the mined-out part of his quarry.

So Bennett added a waste disposal component to his kitty litter business. At first, it seems that the wastes he took were just to procure the oil he needed for his claystone kilns. After a time, however, Bennett began to charge a fee for hauling off and disposing of toxic wastes. As you might expect, this part of the history of the landfill is shrouded in mystery. Nobody really knows how just how much waste Bennett actually disposed of, what kinds of waste, or even where they were buried. But Bennett was quick to see the economic potential of the waste disposal business, and he applied for and received an industrial waste disposal permit. The Pinewood Landfill was born.

One thing is certain about the early history of the Pinewood Landfill—it was a distinctly low-tech operation. The waste drums were rolled to the edge of the quarry, where the waste was dumped in and then mixed with as much processed claystone as was needed to "solidify" the waste. The actual mixing in the pits was done by laborers using nothing more complicated than canoe paddles. Finally, the waste was covered over by whatever fill happened to be available. Because the claystone was considered to be completely impermeable, no further containment was judged to be necessary. All of this, incidentally, was perfectly legal at the time.

In any case, the waste disposal part of Bennett's business soon outstripped the kitty litter-manufacturing part. In fact, he stopped making kitty litter altogether and concentrated on this new and more profitable line of work. The idea of using the claystone as a repository for hazardous waste was so appealing that Bennett was bought out within a year by the Services Corporation of America (SCA). Nobody knows just how much money Bennett made in the deal, but it is a safe guess that his faith in claystone as a commercial commodity was well rewarded.

For the new owners of the Pinewood facility, the timing for acquiring a hazardous waste landfill couldn't have been better. In response to the new environmental movement, Congress passed the Resource Conservation and Recovery Act (RCRA) in 1976. RCRA effectively ended the uncontrolled disposal of hazardous wastes. This meant that opening new landfills—particularly ones that could accept hazardous wastes such as PCBs, dioxins, pesticide residues, heavy metals, and the like—was becoming extremely difficult. At that time, however, South Carolina was anxious to attract industry—any industry—into the state. So, in 1978, Bennett's disposal permit was transferred to SCA, and reissued in 1979. The metamorphosis of kitty litter quarry into hazardous waste landfill was complete.

At this time, the Pinewood facility was legally sanctioned, but just barely. Bennett's original permit had predated RCRA and was not designed to cover a large-scale disposal facility. SCA therefore initiated the lengthy, involved process of obtaining an operating permit, and was granted an interim authorization to continue its disposal activities.

Obtaining a final operating permit for the site largely depended on convincing the regulators at the South Carolina Department of Health and Environmental Control (DHEC) that wastes would be confined within the landfill. It was at this stage that the low-tech mix-and-dump method Bennett and SCA had used began to acquire considerably more sophistication.

It is hard to know just how long it took, or just what stages it went through, but the final design of the waste containment system was anything but low-tech (figure 20.1). The landfill was divided into cells individually excavated into the claystone and separated by berms. Instead of the claystone acting as the only barrier to waste migration, a series of plastic liners separated by compacted clays were built on the bottom of each cell. In addition, a layer of pea gravel was placed on top of the second plastic liner. This layer of pea gravel sloped toward a sump at the end of each cell and was intended to collect any liquid leaking from above. It was on top of this combination isolation–leak collection system that the actual waste was buried. After a cell was filled with waste, the top was sealed by liners and clays to keep rainwater from leaking in.

It is hard to imagine a more comprehensive waste containment system than the one developed and now in use at Pinewood. The whole system is designed to be redundant. If the waste leaked at all, it would be collected by the pea gravel leachate collection system. If that didn't work, the waste would have to pass through two plastic liners and two layers of clay just to reach the claystone. Finally, the waste would have to pass through the claystone in order to get to the ground water. The odds of all of these multiple containment systems being breached, allowing a leak to reach the surrounding ground water, seemed very slim.

Because all of the developmental history is proprietary, it is difficult to track down the origin of this impressive technology. The

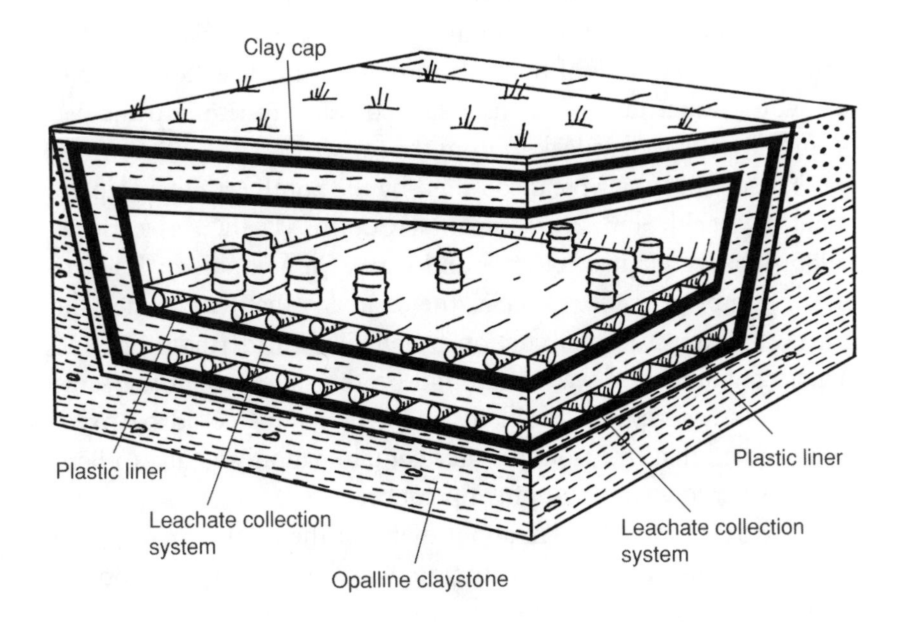

Figure 20.1 Design of the hazardous waste containment cells at the Pinewood Landfill.

timing, however, is fairly well known. In the early 1980s, the low-tech dump-and-mix technology was predominant. It wasn't until 1984 or 1985 that the more sophisticated waste isolation technology was fully in place. So what happened in the meantime?

One thing is clear. The development of the high-tech waste isolation system—of which the landfill owners are now so proud—included the normal process of trial and error. Beginning in the early 1980s, waste disposal was concentrated into different sections, which were in turn divided into individual cells. The integrity of the first section built, conveniently called "Section 1," seems to have suffered from the inexperience of the landfill operators. For one thing, they managed to perforate the claystone while digging the trenches, thus losing any containment offered by the claystone. For another, they cut one of the cells into an area that is now actually below the water table. This has allowed water to rise into the cell from below, as well as leak in from the top.

These problems were discovered by the South Carolina Department of Health and Environmental Control (DHEC) as early as 1985, when their inspectors found excessive amounts of leachate in the sumps of Section 1. GSX, the company that now owned the site, was given a formal citation by DHEC. Furthermore, ground water contamination was also found on the site at about the same time. GSX insisted the problems with the sumps could be fixed, and claimed that the ground water contamination was due to Bennett's earlier low-tech dumping operation.

GSX was probably right about that, but the local citizenry, who had long been unhappy with the growing hazardous waste operation, were in no mood to listen. As far as they were concerned, ground water was being contaminated and their water supply was in jeopardy. All of this was exacerbated by the Pinewood Landfill's being only a few hundred yards from Lake Marion, one of the most popular fishing spots in the state and the headwater for numerous municipal water supply systems in the Santee-Cooper River Basin.

Citizens began to come forward with tales of strange sicknesses. Fish caught in Lake Marion were reported to be covered with cancerous sores. A local citizen's group called CASE (Citizens Asking for a Safe Environment) was formed; its demonstrators went so far as to lie down in the road to block trucks from entering the facility. Municipal ground water users as far away as Myrtle Beach (a good seventy miles away as the crow flies) expressed concern.

There was, in short, general panic.

And it is hard to blame the people for being scared. The more that became known about the landfill, the more there seemed to be grounds for concern. As additional monitoring wells were drilled, it became increasingly clear that the supposedly simple geology at the site was not so simple after all. In some places, for example, the claystone was more than 100 feet thick, whereas in others it was absent altogether. Because the claystone had been touted as the ultimate barrier to contaminant migration, the apparent fact that the claystone wasn't always there was certainly worrisome.

But that wasn't all. Underlying the claystone was about fifty feet of gray sediments, named the Sawdust Landing member of the Rhems formation, that nobody seemed to be able to make any sense out of. In some places, Sawdust Landing sediments were dense clays—perfect for further containment of any leaking contamination. In others, the Sawdust Landing consisted of coarse sandy material—perfect for transporting any contaminants that happened to reach it. Dozens of wells were drilled into the Sawdust Landing member in an attempt to figure out what was going on, but this only served to deepen the confusion. It was possible, for example, to drill a well and encounter a sandy layer at a depth of about fifty feet. If the driller moved over ten feet and drilled another well, the sand might be missing altogether. Drilling so many wells was actually adding to the confusion, not helping to resolve it. What could possibly be causing all of this strangeness?

While these problems were perplexing, they weren't considered serious enough to interfere with the flow of hazardous wastes to the landfill. By 1986, about 100,000 tons of various toxic and hazardous chemicals had been stowed in the Pinewood facility.

Meanwhile, the Santee-Cooper authorities, the people responsible for Lake Marion, were getting worried. The sheer amount of toxic chemicals accumulating at the landfill was scary just by itself, and the increasing confusion about the geology of the site didn't help allay their fears. It was time, they decided, for a comprehensive investigation of the hydrogeology of the site.

At this point, the GSX people and the Santee-Cooper authorities had a stroke of luck. A geologist named Dave Prowell happened to be doing some geologic mapping in South Carolina at the time, and was interested in adding the area around Pinewood to his maps. Prowell started work in 1986, and in just over a year he began to resolve the mysteries surrounding the site.

According to Prowell, the Sawdust Landing was deposited in a fluvial environment, that is, the sediments were carried in by rivers and deposited as the rivers moved back and forth on their flood plains. This immediately explained why some places in the Sawdust Landing were sandy, while others, just a few feet away, were clayey. Rivers naturally separate coarse- and fine-grained sediments and deposit them in different places, depending on how fast the current happens to be flowing. That is just the way fluvial sediments are.

It seems that after deposition of the Sawdust Landing member was complete, there was a sudden rise in sea level. The sea inundated the area surrounding the future landfill, but in an unusual way: a series of barrier islands to the east greatly restricted the flow of seawater. The area probably looked something like Long Island Sound does today. In any case, fine-grained sediments accumulated to a depth of about 200 feet. By the end of the Paleocene epoch (about 58 million years ago), the deposition of the claystone was complete.

But a lot can happen in 58 million years, and, in the case of the claystone, a lot did. Sea levels continued to bounce up and down, but on the whole, the area surrounding the future landfill lay exposed at the surface. During this time of exposure, the ancestral Wateree River gradually eroded through the claystone. This erosion was most pronounced, of course, near the river and in places completely removed the claystone. It was this erosion that caused the apparently inexplicable thinning of the claystone near Lake Marion. The Wateree River did not just erode material, however. In places, the river deposited layers of sand and silt on top of the claystone.

By 1990, thanks largely to the efforts of Dave Prowell, the geologic understanding of the Pinewood site was considerably clearer that it had been in 1978. Even though the distribution of sands in the Sawdust Landing member remained vexingly complex and unpredictable, the reasons for this unpredictability were now known. Furthermore, buried deep within all of this geologic complexity was something hydrologically very simple—the site is located adjacent to a classic ground water discharge area (figure 20.2).

A "ground water discharge area" is a place where water percolates *out* of the aquifers (discharges), instead of the other way around. If you drill wells into successively deeper aquifers near such an area, the water levels in the wells become *higher*. This means, of course, that ground water is flowing *upward*, and in this case, seeping into Lake Marion—which is why the Wateree River and Lake Marion are there in the first place. So, even if contamination were to leak out of the landfill, it would not be drawn downward into the underlying aquifers, it would seep laterally and eventually enter the lake.

So we are left with a classic good news-bad news situation. The good news is that even if the high-tech engineered barriers are

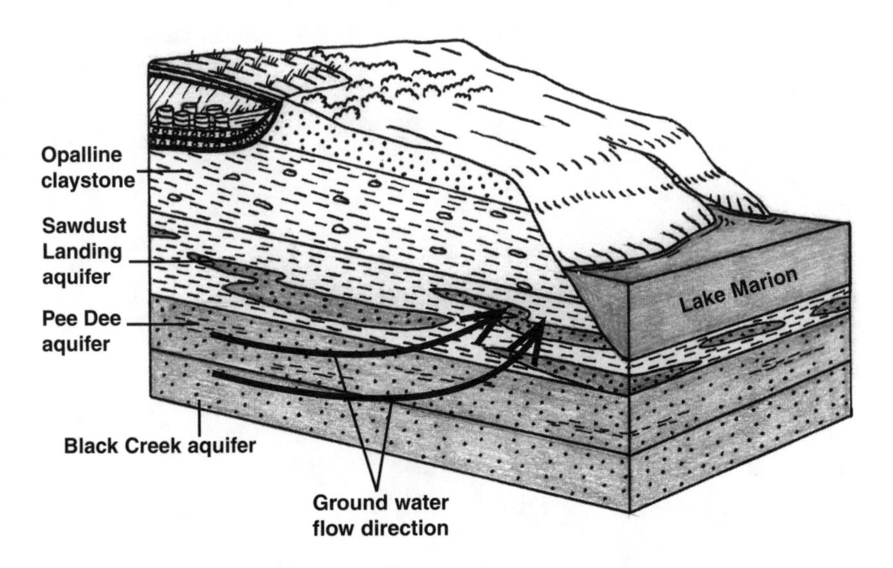

Figure 20.2 Geology and ground water flow patterns in the aquifers underlying the Pinewood Landfill.

breached, extensive ground water contamination from this site is highly unlikely. Myrtle Beach can certainly breathe easy. The bad news, on the other hand, is that if any significant leakage occurs, there is a good chance the contamination is going to end up in Lake Marion.

THE PINEWOOD SECURE LANDFILL is a microcosm of the revolution that occurred in the hazardous waste industry throughout the 1970s and 1980s. It is still entirely debatable whether the safeguards built into landfills like Pinewood are sufficient to maintain waste isolation over the long run. What is not debatable is that such operations are a huge improvement over what passed for "proper" waste disposal in the very recent past. The question is, are they enough?

Only time will tell.

Chapter Twenty-One

The Curse of Sisyphus

THE REVOLUTION IN HAZARDOUS WASTE DISPOSAL PRACTICES that occurred in the 1970s and 1980s—as illustrated by the story of the Pinewood Landfill—has gone a long way toward preventing new instances of ground water contamination. But many, many contamination events occurred long before this revolution had taken hold. As a result, we have been left with an extensive legacy of ground water contamination problems.

Since we can't turn back the clock, we need to be able to deal with—and hopefully clean up—this contaminated ground water. But even though there are a number of technologies available to clean up contaminated ground water, they all take lots of time. Part of the reason is that the technologies are crude and still under development. But the other, more important part is that the hidden, isolated nature of ground water systems makes them inherently difficult to clean up.

Consider, for example, the Savannah River Site in South Carolina.

THE BEGINNING OF THE COLD WAR in the late 1940s saw a frantic scramble in the United States to set up the industrial infrastructure necessary to produce nuclear weapons. All over the United States, plants sprang up to help build and assemble the specialized components of atomic, and later hydrogen, bombs. The places chosen for this new growth industry were not, in general, particularly well known. Hanford, Washington; Miamisburg, Ohio; Oak Ridge, Tennessee; Rocky Flats, Colorado; and the Savannah River Site, South Carolina were just some of the cogs in the nuclear arms production machine that swept into high gear around 1950.

Most of these plants were set up in a tremendous rush, with secrecy, and with the conviction that this was what was best for the country. Rural South Carolina, with its relative isolation was just the sort of place to make nuclear materials and assemble atomic weapons. So, in 1950, the Atomic Energy Commission—the predecessor of the Department of Energy—acquired 300 square miles of land near the Savannah River. The site was to be used to manufacture enriched plutonium and tritium—the radioactive components of thermonuclear weapons.

Eventually, four nuclear reactors were built to produce weapons-grade nuclear materials. An assortment of machine and tool shops, manufacturing facilities, chemical synthesis and separation lines were built to support the facility. By 1960, these facilities were working virtually around the clock to support the nuclear arms race with the Soviet Union. The word "race" is particularly appropriate because these facilities were managed with one and only one goal in mind—to produce the materials needed as fast as possible and the devil take everything else.

One recurring problem at all of these weapons sites were the wastes being generated. The reactors produced high- and low-level radioactive wastes, the chemical production facilities produced an assortment of noxious wastes, and the day-to-day operation of machine tool facilities also generated chemical wastes. So how were these wastes handled? Each weapons facility had its own way of solving the problem, but they shared one characteristic—whatever got rid of the wastes in a hurry was the method of choice.

At the Savannah River Site (SRS), there were some remarkably easy ways to get rid of chemical and low-level radioactive liquid wastes. Located in the Upper Coastal Plain, the site is underlain by relatively coarse-grained sands and gravels that were deposited by rivers some 70 million years ago. Because these sands and gravels were so permeable, the site managers soon found that all they had to do to get rid of liquid wastes of most any kind was pour them into unlined seepage ponds and let them sit, where, after a few days, they would either evaporate or infiltrate into the ground and be gone. Fortunately, even these time-and-cost-driven managers had the sense not to dispose of high-level radioactive wastes—the fission products of uranium and plutonium—in this way, but instead to store them in underground storage tanks. Nevertheless, hundreds of millions of gallons of low-level radioactive wastes, tritium mostly, were simply dumped into unlined pits. Along with the radioactive wastes went millions of gallons of assorted chemical wastes.

Some of the least noxious wastes generated at the site were chlorinated solvents. For example, the M Area of the Savannah River Site was devoted to fabricating specialized components of nuclear weapons. Because it wasn't feasible to machine actively radioactive materials into fuel elements that made up the core of nuclear weapons, some components were fabricated out of nonradioactive materials and then irradiated in the SRS reactors. And because, to machine these elements, it was necessary to clean them of all traces of grease, various chlorinated solvents were used in great quantity.

One of the most commonly used degreasing solvents was trichloroethene (TCE). This compound consists of two carbon atoms joined by a double bond, with three attached chlorine atoms and one hydrogen atom. Because the electrical charges of the chlorine and hydrogen atoms are so evenly distributed around the molecule, TCE efficiently dissolves greases and fats. It is because of this very property that TCE is also used to dry-clean clothes.

Between 1952 and 1962, TCE was the solvent of choice in the machine shops. Between 1962 and 1971, it was perchloroethene (PCE), and after 1971, trichloroethane (TCA). From 1952 to 1982, an astonishing 13 million pounds of these solvents were used in the M Area alone. As wastes, they were promptly flushed into the M Area settling basin, where about half evaporated and the other half seeped into the sandy soil and obligingly disappeared.

By the time the practice was discontinued, however, the M Area had developed one of the largest plumes of TCE-, PCE-, and TCA-contaminated ground water in the world. Fortunately, this plume was not moving toward any populated areas outside of the SRS. But the fact that it was less than ten miles from New Ellenton—a town that relied on ground water for its water supply—caused a political crisis. Within the space of a few years, the major business at the M Area went from dumping solvents into the ground to trying to extract those very same solvents back out of the ground.

So how do you remove 5 or 6 million pounds of TCE, PCE, and TCA from an aquifer? It soon became evident that nobody had any idea how to do this. One obvious way to do it was to drill wells into the plume, pump out the contaminated water, and treat it to remove the wastes. A "pump and treat" system was the first technology tried at the SRS, and between 1984 and 1990, it removed about 230,000 pounds of solvents—a tiny fraction of the 5 or 6 millions pounds of contaminants. At that rate, the system would have to run for over a hundred years to remove the contaminants from the aquifer.

Government bureaucrats are not, on average, a patient bunch. The dawdling pace of cleanup would have to be speeded up, came the decree from Washington. That was easy to say, but somewhat harder to do. However, necessity and the availability of money are the mother of invention. Because the Savannah River Site had very deep pockets, the SRS managers soon had a variety of proposals on the table designed to speed the cleanup process.

And some of the ideas were pretty good. A major limitation of conventional pump-and-treat systems was that ordinary vertical extraction wells could penetrate only a limited portion of the contaminated aquifer. A well drilled vertically downward into a contaminated zone fifty feet thick, for example, could intercept only fifty feet of the zone. But if a well could be drilled downward and then *sideways* into that same zone, it could intercept hundreds of feet of contaminated aquifer. "Horizontal wells," as they were called, might vastly increase the efficiency of the pump-and-treat system.

Another major limitation, however, had to do with pump and treat itself. Processing and treating the huge amounts of contaminated ground water in aboveground treatment facilities was difficult and expensive. Was there some way to separate the solvents from the water underground? Yes, there was. TCE, PCE, and TCA are all fairly volatile, which is to say, they will evaporate much more readily than water. If air could be blown into the aquifer, the solvents could be volatilized. Furthermore, if this solvent-laden air could then be extracted and incinerated, the water treatment leg of pump and treat could be bypassed altogether. This technology—called "in situ air stripping"—also had the potential to vastly speed up contaminant removal at the SRS.

There was a third possibility as well. TCE, PCE, and TCA are not biodegradable under ordinary aerobic conditions. Most bacteria cannot, for example, use these compounds as sources of carbon or energy. However, one class of bacteria—the methane-oxidizing bacteria—produce an enzyme, methane monooxygenase (MMO), designed to insert oxygen into methane. Because MMO is not a very picky enzyme, in addition to oxidizing methane, it will just as happily oxidize TCE, PCE, and TCA. And if you injected both air and methane into the aquifer, you could develop a large population of these methane-oxidizing, contaminant-eating bacteria.

So, between horizontal wells, in situ air stripping, and methane-oxidizing bacteria ("bioremediation"), there were three possible ways to speed up removal of the solvents from the aquifer underlying the M Area. The purely scientific way of evaluating each of these possible methods would have been to apply them in turn, so that the relative efficiency of each could be determined. The site managers would have none of that, however. They were in a hurry. So they decided to combine all three methods—horizontal wells, in situ air stripping, and bioremediation—into one demonstration project.

In 1988, two horizontal wells were installed in a portion of the plume. The original idea was to drill one well about twenty feet below the water table (where most of the contamination was) and to drill the other one parallel to the first, but above the water table. Unfortunately, the drillers were off by a little, and the wells weren't quite parallel. But, considering the technical difficulties involved, they didn't do too badly.

Finally, in 1992, the full system—horizontal wells, in situ air stripping, and methane-oxidizing bioremediation—was ready to be tried. First, the upper horizontal well was used to extract air from above the contaminated aquifer. This was done mainly to test the efficiency of the electrically heated catalytic oxidation system, which had the job of destroying the solvents before they could be spewed out into the air. Second, air was injected into the lower injection well while the upper well continued to extract contaminants. This went on for about 220 days and led to the extraction of about 6,000 pounds of solvents. Finally, methane was added to the stream of injected air in order to stimulate the growth of MMO-producing, methane-oxidizing bacteria.

As you might expect, making all of this come to pass was an expensive undertaking. By 1992, the project had cost about 5 million dollars. Naturally enough, the question was how well had the system worked. In just over six years of operation, the conventional pump-and-treat system had removed TCE at an average rate of 105 pounds of solvents per day. By contrast, this experimental system—

including the horizontal wells, the air stripping, and the MMO-producing bacteria—managed a removal rate of about 50 or 60 pounds per day. Thus, even with virtually unlimited funds, making use of the most sophisticated and advanced technology available, and addressing fairly accessible contamination, the experimental system had achieved only half the removal of the conventional system. In all fairness, the conventional pump-and-treat system was much larger than the experimental system, so the numbers can't really be compared. But the bottom line is that it will still take decades or centuries to clean up these aquifers.

THERE IS A STORY IN GREEK MYTHOLOGY that aptly sums up the problems inherent in trying to clean up severe instances of ground water contamination. It concerns a onetime king of Corinth named Sisyphus, who happened to observe Zeus, the king of the gods, abducting a young maiden named Aegina. Zeus took Aegina to a nearby island, where he could, as was his custom, debauch her at his leisure. When the girl's worried father came to look for her, Sisyphus identified both the abductor and his motive. This was bad judgment. Zeus flew into a rage, drove off the father with a salvo of thunderbolts, and then went looking for the unfortunate Sisyphus. For his effrontery, Zeus charged Sisyphus with the task of pushing a huge boulder to the top of a hill—a difficult but a doable task. The catch was that Zeus had enchanted the boulder so that each time Sisyphus almost got it to the top, it would escape his grasp and roll back to the bottom of the hill. Sisyphus was condemned to spend eternity trying, but not quite succeeding, to accomplish his assigned task.

It was slowly becoming clear that the Savannah River Site had been charged with just such a task. No matter how hard they tried, no matter how much money they spent or what technology they used, the SRS was doomed to work and work and work, without ever being able to quite finish the job.

It is the modern Curse of Sisyphus.

Chapter Twenty-Two

A Model of Uncertainty

THE FACT THAT CLEANING UP GROUND WATER is inherently difficult and time-consuming can be very worrisome to people living near the contamination. Because many sites, like the Savannah River Site, cannot be cleaned up quickly with even the best available technology, it is natural to wonder just where the contaminated ground water will move in the future and whether it is likely to affect neighboring water supplies. Over the years, mathematical models have been developed to peer into the hydrological future and predict just where and how fast the contaminated ground water is moving. But like any technology, developing such models involved a good bit of trial and error.

Take, for example, the Snake River Plain aquifer of Idaho.

IT ALL BEGAN DURING WORLD WAR II. An important part of the Pacific Fleet at that time were the huge battleships and their 16-inch guns (that is, the diameter of the shells they fired was 16 inches!). Because of the incredible stresses generated by firing such huge shells, however, the guns had to be rebored fairly frequently. And after they had been rebored, they needed to be thoroughly tested before they were sent back into combat. Someone somewhere in the Department of the Navy decided that the remote desert on the Snake River Plain was a perfect site for this testing. That is how what is now called the Idaho National Engineering Laboratory (INEL) came into being.

After the war, the Navy and the Atomic Energy Commission converted the INEL into a facility for building and testing nuclear reactors to be used on ships. The reactor for the Navy's first nuclear submarine, the *Nautilus*, was built and tested at the INEL. But the research at the INEL was not just confined to military uses of nuclear energy. It happens that the first reactor capable of producing electricity from nuclear fission was built and tested at the INEL. Later, as the nuclear age got into high gear, facilities for reprocessing spent fuel rods from the Navy's reactors were added to the INEL's capabilities. Nuclear reactors generate a number of radioactive by-products ("radionuclides") such as tritium, strontium-90, cesium-137, cobalt-60, plutonium-238, -239, and -240, and iodine-129. By 1953, the INEL was producing around 200 million gallons of this radionuclide-containing water per year. But, happily for the site operators, disposing of liquid wastes at the INEL proved to be easy. This is because the Snake River Plain is underlain by one of the most unusual ground water systems is the world (figure 22.1).

In the recent geologic past, the area surrounding the INEL has experienced intensive volcanic activity. The volcanoes that developed produced a very liquid lava that, after emerging from the vents, spread out over the Snake River Plain like floodwaters. In the 1.6 million years that this volcanic activity has been going on,

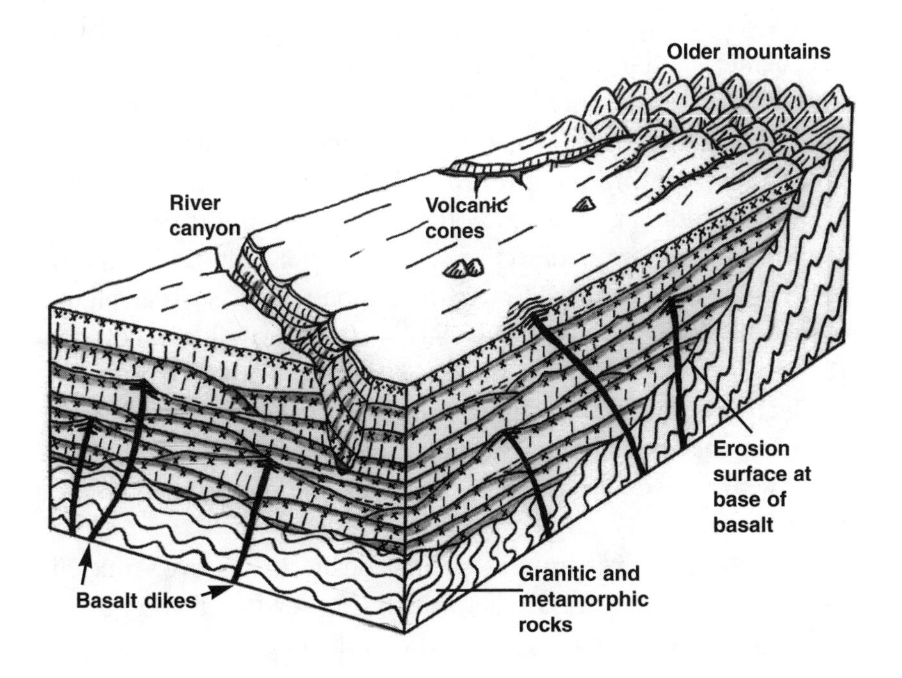

Figure 22.1 Basalt aquifers underlying the Idaho National Engineering Laboratory (modified from Heath, 1984).

there have been at least 120 different episodes of these lava "floods." A bit of arithmetic indicates that these lava floods took place, on the average, every 13,000 years or so, with the last episode occurring about 2,000 years ago. In between the episodes of volcanic activity, lakes developed on the plain, so that the INEL site is underlain by a series of lava flows, which range in thickness from 5 to 50 feet, interbedded with somewhat thinner beds of lake sediment.

Rock from this kind of lava (called "basalt") is full of bubbles (called "vesicles") that form as gas escapes the cooling rock, and full of cracks (called "columnar joints") that form as the rock cools and shrinks. All of this has led to an extremely permeable and productive aquifer system—which was perfect for the disposal of liquid chemical and low-level radioactive wastes.

So when low-level radioactive wastes began to be generated at the INEL in the early 1950s, these wastes were summarily disposed of by putting them down a 580-foot well into the Snake River Plain aquifer, a practice that continued until 1984. This method of disposing radioactive waste was not quite as foolhardy as it might seem at first. Most of the radionuclides, such as plutonium-238, -239, and -240, iodine-129, strontium-90, and cesium-137, are not particularly soluble, so that once injected into the aquifer, they would bind strongly to the rocks and not move very far. Even the tritium, present as "heavy water," which migrated freely with flowing ground water, was not too worrisome, at least not at first.

Being radioactive, tritium spontaneously decays over time. It has a half-life of only 12.3 years, which is to say that if you put 10 curies (a standard unit for measuring radiation) in an aquifer in 1955, only 5 curies would remain in 1967. So, the reasoning went, go ahead and inject the stuff into the aquifer, and by the time the tritium reaches the boundaries of the INEL site, it will have completely decayed into nonradioactive helium.

All of this seemed reasonable enough, except for one thing. Because the Snake River basalts are so permeable, ground water moves at the unusually fast rate of 10 or 20 feet per day. Furthermore, ground water at the INEL moves south and west, eventually discharging near the Snake River in a series of large springs that serve as an important source of water for agriculture and commercial fisheries. If the tritium was ever to reach these springs, there could be real trouble.

Thus, even though it seemed reasonable the tritium would decay long before it left the INEL site, and certainly long before it reached the springs, there was always a nagging doubt that it might not. It was this uncertainty that led the managers of the

INEL to look for a way of forecasting how far the tritium would eventually go.

On the surface of it, this looks like it shouldn't be too hard to determine. After all, if you inject, say, 10,000 curies of tritium into the ground water, and the ground water is moving 20 feet per day, then right away you can calculate that the tritium will move 7,300 feet, or 1.4 miles, in just one year. Because the plant boundary is only about 9 miles away, it will only take about six and a half years for the tritium to get there. In that short a time, most of the tritium (about 7,500 curies) will remain undecayed, and it seems certain that most will therefore leave the site.

But, it isn't that simple. For one thing, the tritium was added not in one gigantic slug but more continuously, at a rate of about 900 curies per year. For another, once the tritium mixed with the ground water, it would be continuously diluted as it moved southward. And finally, as the tritium moved, it would be dispersed by the flowing ground water. These three processes, all of which tend to decrease concentrations of tritium, turn this into a problem that cannot be solved by simple arithmetic. It just gets too complicated to try and add up all of the sources (injection) and "sinks" (radioactive decay, dilution, dispersion) for tritium.

Which is not to say the problem can't be solved. The mathematical trick is to add up these sources in terms of the *changes* in tritium concentrations. If you pick any part of the aquifer to serve as a reference, then the change in tritium concentrations over time must equal the changes due to (1) the amount of hydrodynamic dispersion of the solute, (2) the amount of advective mixing with nonradioactive water, (3) the amount of tritium adsorbed onto aquifer material, and (4) the amount of radioactive decay. In short, you end up with a rather complicated equation.

If you have a computer handy, this equation can be solved fairly easily. Because such computers were not available in the 1950s and early 1960s, there was no real way to know just how far the radioactive wastes injected at that time would migrate in the aquifer. By the late 1960s, however, computers did become available, and hydrogeologists were soon putting them to use to "predict" the future behavior of the INEL tritium plume.

It is very easy to be impressed by the elegant mathematics embodied in this approach. However, if you look closely, you will see several very obvious uncertainties. For example, a whole slew of parameters describing tritium adsorption, dispersion, ground water velocity, and radioactive decay are needed. The problem is, there is no easy way to measure these parameters in an aquifer four or five hundred feet below land surface. Furthermore, ground water is not going to have the same velocity everywhere. How can you measure that?

The point is, even though hydrologists now had an equation that described precisely how tritium would behave, and even though they had a computer that could solve the equation, a good bit of uncertainty remained. Without reliable ways of measuring all the needed parameters, educated guesses would have to suffice. And the liberal use of educated guesses carries with it the baggage of uncertainty.

In the early 1970s, an enterprising hydrogeologist named J. B. Robertson used this equation in order to predict how far and how fast the tritium would migrate. Robertson soon bumped into the problem of how to estimate the parameters needed by the model. It was a classic example of a hydrological catch-22. He needed to know values for dispersion, velocity, adsorption, and radioactive decay, but the only parameter he knew with any certainty at all was the half-life of tritium. So Robertson went about doing what any good scientist would do in this situation.

Figuring that it was better to overestimate how far the tritium would migrate, Robertson set about using high estimates for the dispersion and velocity parameters (which "speeded up" simulated rates of tritium transport) and low estimates for the adsorption parameters (which "slowed down" transport). Cranking up his computer, he boldly predicted the distribution of tritium in the aquifer for the years 1980 and 2000. In doing so, Robertson was going out on the same limb meteorologists go out on every day—he made his best estimates and took his best shot.

As you might expect, actual concentrations of tritium in the ground water were routinely monitored at the INEL. So, once 1980 rolled around, it was possible to go out, measure how far the tritium had actually moved, and compare it to what Robertson's model had predicted. It is tempting to view this sort of comparison solely in terms of whether Robertson was right or wrong. But there is actually more to it than that.

When Robertson built his model, he knew that there was considerable uncertainty in the dispersion, velocity, and adsorption parameters he had used, largely because there was no way to actually measure them. By 1980, this situation was largely unchanged: there was still no method for directly measuring these parameters. But the very fact that Robertson had made his predictions opened a brand-new window. By comparing the observed behavior of the tritium to what Robertson had predicted, it would be possible to make indirect estimates of the parameter values. In other words, by knowing where Robertson had been wrong, his successors could make new estimates that were substantially better.

For example, Robertson's model predicted that, in 1980, tritium concentrations three miles south of the disposal well would be about 40 picocuries per milliliter (pCi/mL) of water (a picocurie is one-trillionth of a curie). The actual measured concentration at this location was closer to 10 pCi/mL. In other words, Robertson's prediction was off by a factor of 3. Other predictions made by the model didn't quite come to pass either. The model predicted that water containing about 2 pCi/mL of tritium would reach the INEL boundary by 1980. The actual measured concentrations of tritium at this location were below 0.5 pCi/mL.

So was Robertson just flat wrong? Not really. Because he knew he was going to probably be wrong anyway, he had tried very hard to overestimate and had succeeded admirably. Robertson's successors now had concrete evidence that his parameters were overestimates. In particular, the value for the velocity of ground water was too high. Rather than 20 feet per day, as Robertson had assumed, it now seemed clear that ground water was moving at the somewhat more sedate speed of 10 feet per day.

YOU HAVE TO FEEL A CERTAIN SYMPATHY for people who spend their time working on predictive models—either hydrological models like Robertson's, or meteorological models that predict the weather. Because of the inherent uncertainties involved in the systems, these people know that their predictions will be wrong at least to some extent. Furthermore, it is only by carefully analyzing just how wrong these predictions are that these uncertainties can be reduced.

In a strange and convoluted way, the success of a model is often measured by the magnitude of its failure.

Chapter Twenty-Three

The Best Intentions

It can certainly be argued that, if the people running the Savannah River Site or the Idaho National Engineering Laboratory had been more conscientious about disposing of their wastes, they could have avoided many of the resulting problems. Conscientiousness, however, isn't always enough. It is just a sad fact that well-intentioned people sometimes make bad situations substantially worse. This happens often enough in all aspects of human life, but it is particularly common in instances of ground water contamination. When dealing in the realm of the uncertain, the unclear, and the improbable, good intentions—particularly, good intentions mixed with a lack of knowledge—can cause real harm.

And there is no better example of this than what happened at Hanahan, South Carolina.

HANAHAN IS NOW A SMALL BEDROOM COMMUNITY located within the urban sprawl of Charleston, but back in World War II, it was well out into the countryside. Because of its isolated location, and because of the proximity of the Charleston Naval Base and the Charleston Air Force Base, Hanahan was deemed to be a suitable location for a fuel storage facility. During the war and in latter years, the Hanahan Defense Fuel Supply Center (DFSC) came into being. The DFSC today consists of six huge aboveground storage tanks, each of which holds millions of gallons of fuel.

Throughout the 1950s, the DFSC stored and supplied fuel for the Navy and Air Force. They stored aviation gasoline, an ultra-high-octane gasoline used in propeller-driven aircraft, leaded gasoline, a concoction called "MO-gas" used in armored vehicles, and jet fuel. In the 1960s and 1970s, much of the storage capacity was taken up by a jet fuel called "JP-4." This was largely due to the fuel needs of the nearby Charleston Air Force Base.

The 1960s and 1970s saw a tremendous growth in the population around Charleston and Hanahan. Gradually, the suburban sprawl enveloped even the unpromising marshy land surrounding the DFSC. The marshes were drained, and houses began to be built. Eventually, a neighborhood grew right up to the fence line of the DFSC. Most of the people living in the neighborhood were used to the military and tended to ignore the presence of the huge storage tanks. If anybody had any concerns at all, they had more to do with the potential fire hazard then anything else.

As it happened, however, the real hazard had little to do with fire. In 1975, a routine fuel inventory turned up an apparent deficit in Tank 1. The people running the DFSC at the time were conscientious folks, so they immediately transferred all of the JP-4 fuel from Tank 1 to other tanks and began looking for a leak. This was a difficult task and involved excavating the underground pipes that led to and from the tank.

Soon, however, the leak was found in a pipe buried several feet below the ground. As such, it had not been evident for some time, and a huge amount of fuel had escaped and soaked into the sandy soil. The fuel had migrated downward to the water table, which was about fifteen feet below land surface, and because JP-4 is lighter than water, it had spread out like a lens (figure 23.1). The actual amount of fuel lost was estimated to be 83,000 gallons.

The Defense Logistics Agency, the group responsible for the DFSC, was aghast that a leak of this magnitude had occurred. The fact that it had happened in an area so close to where people were living made the situation even worse. The fuel could seep into drains or sewage lines, vaporize, and cause a substantial explosion hazard. Furthermore, toxic chemicals like benzene could be leached from the fuel and contaminate well water. The spill must

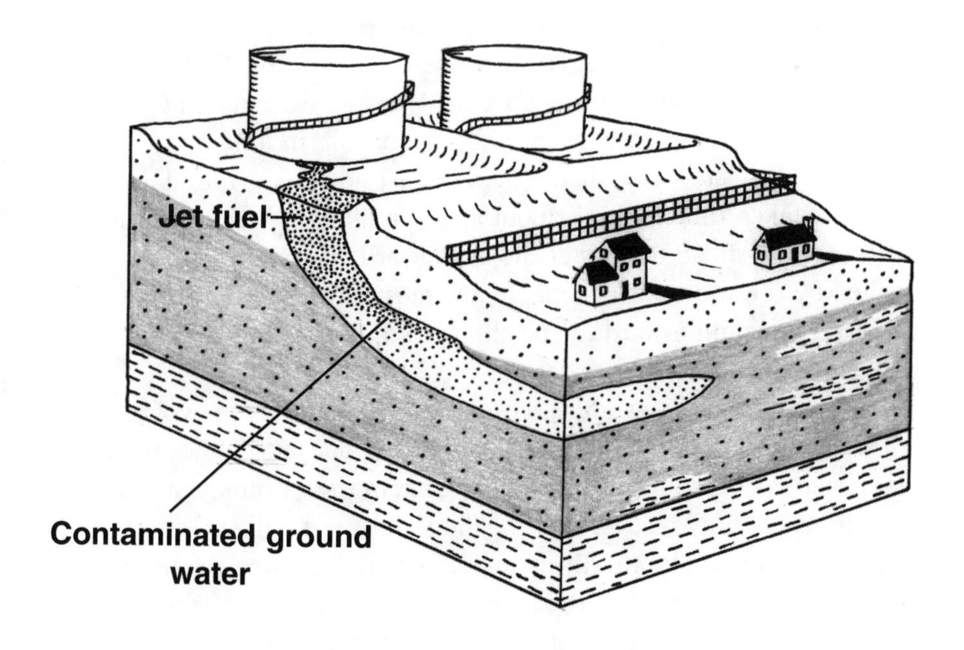

Figure 23.1 Jet fuel contamination of ground water at the DFSP tank farm in Hanahan, South Carolina.

be cleaned up immediately, came the decree from above, and no cost was to be spared in the cleanup effort.

So far, you can't fault either the actions or the intentions of the DFSC people. After all, it was their conscientiousness that located the leak in the first place. Furthermore, once the leak was found, you can't fault the zeal with which they pursued the cleanup. But it was this very zeal to clean up the spill that, paradoxically, made the situation substantially worse.

The engineers charged with cleaning up the spilled fuel decided to drill a large number of wells, and to vigorously pump both the ground water and the JP-4 out. The fuel-water mixture would then be routed to an oil-water separator, where the JP-4 could be recovered. Drilling rigs were brought in and more than a hundred recovery wells drilled. Pipes were laid to collect the huge amount of water and JP-4 that would be produced, and an oil-water separator installed. Then, without any further delay, the recovery system kicked into high gear. The spill occurred in October, and by December the recovery operation was in full swing.

In retrospect, it is a shame there wasn't any delay. For, if anyone had taken time to think about it they might have realized that, because JP-4 floats on water, its downward migration had been halted by the water table. True, it was most likely spreading laterally and downhill toward the neighborhood, but at a rate of probably only an inch or less per day.

In instances of fuel spills, it is crucial to minimize the volume of aquifer contacted by the fuel. If you take a bucket of sand, pour in 5 gallons of gasoline, and then try to drain the gasoline out, you will probably only recover 3 or 4 gallons. The rest will stick to the grains of sand and be effectively immobilized. So, before the recovery system was turned on, the JP-4 was concentrated in a relatively small area, and a relatively small portion of the aquifer had been impacted.

But as soon as the hundred or so wells began pumping, all of this changed. This massive pumping immediately lowered the water table by thirty feet and smeared JP-4 over the entire aquifer (figure 23.2). Thus, in the space of a few days, the volume of aquifer contaminated by the JP-4 was increased by more than a factor of 10. But that wasn't the worst of it. The increased contact with sediments had effectively adsorbed a much higher portion of the JP-4 than had been the case originally, and some of the fuel had become trapped underneath lenses of clay that were sprinkled throughout the aquifer. Not only had the volume of contaminated aquifer been greatly increased, but the amount of recoverable JP-4 had been greatly reduced.

But, sad to say, nobody noticed any of this at the time. At first, a lot of JP-4 was being recovered, and the engineers were satisfied that they had done the right thing. The only indication that things weren't going absolutely perfectly was that the amount of JP-4 actually recovered at the oil-water separator was less than they had hoped. Of the 83,000 gallons of JP-4 spilled, they recovered only

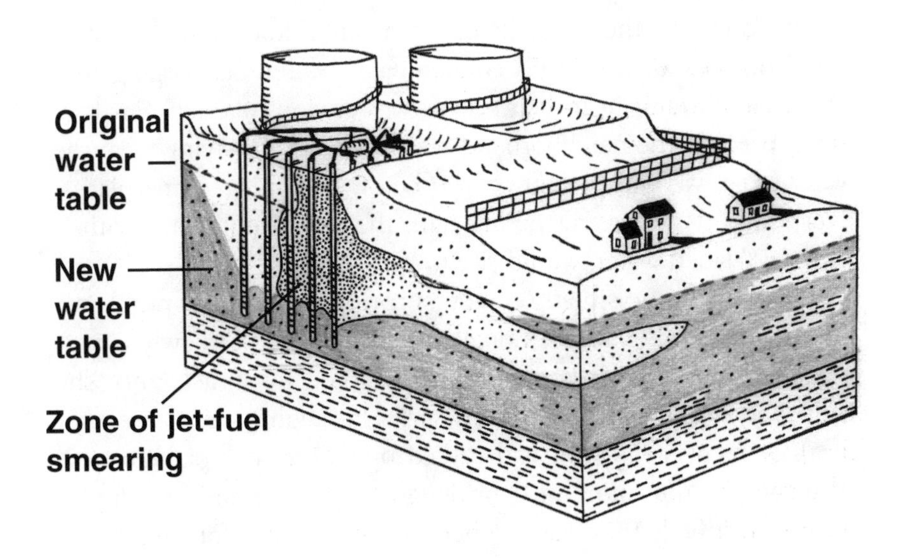

Figure 23.2 Lowering the water level by pumping smeared jet fuel throughout the aquifer.

about 20,000 gallons. After a few weeks, the amount of JP-4 being recovered dropped off substantially. Finally, when it appeared that they had recovered everything they could, the system was turned off. The only mystery was what had happened to the other 63,000 gallons of fuel. Where could it be? It was hopelessly trapped in the aquifer. Each grain of sand was thoroughly coated with a sheen of JP-4, and there were pockets of pure fuel trapped underneath clay lenses scattered though the aquifer.

As soon as the fuel recovery wells stopped pumping, water levels slowly rose again in the newly contaminated aquifer. Ground water again began to flow toward a neighborhood located a few hundred feet north of the site. Now, however, the ground water had to pass through an aquifer that was coated with JP-4 from top to bottom, and as it did, it picked up the soluble components of the fuel, principally benzene, toluene, ethylbenzene, and xylene (commonly referred to as "BTEX") and carried them directly into the neighborhood.

It didn't take the residents of the neighborhood long to notice this. The reek of the BTEX-contaminated water discharging into streams and drains near their houses was obvious enough. Furthermore, because the neighborhood was in such a low-lying area, the water table was only a foot or two below ground in some places. Every time there was a soaking rain, the water table, and with it the BTEX-laden water, would rise to land surface.

This unhappy situation, not surprisingly, caused problems. One of the families whose home was affected by the newly contaminated ground water had a young daughter. By all accounts, she had always been a happy, healthy girl. One night, some time after the JP-4 recovery system had ceased operating, her parents were awakened by the sound of their daughter choking and gasping for breath. Terrified, they rushed her to the hospital for emergency treatment. At the hospital, they were told that their daughter had had an asthmatic attack. It is impossible to know if the attack had been caused by the BTEX fumes seeping into the house, but it is certainly possible. The angry parents, however, were not troubled by any doubts. As far as they were concerned, the contamination had caused their daughter's illness, period. And who knows? Maybe it did.

In any event, the entire matter ended up in court, with the neighborhood residents charging the site operators with gross negligence and asking for about 20 million dollars in damages. The site operators, for their part, were miffed by any suggestion of negligence. After all, they had been forthright about the whole affair and had moved aggressively—and at great cost—to correct the problem before it got worse. There was no way, they thought, that a charge of negligence could stick. So the lawsuit began its slow progression through the legal system.

At first, the site operators were confident that they could win in court. However, as time went by and as the hydrogeology of the site was studied in some depth, it became clear that their actions—while well-intentioned—had exacerbated the contamination problems associated with the spill. Also, it became evident that the residents' case had weaknesses as well. After all, the site operators had simply taken the considered advice of their engineers. Negligence would be difficult to prove.

In the end, the residents settled with the site operators for about 2 million dollars. It would have cost the operators about that much money to litigate the case anyway, so they offered it as a settlement. Much of the money went to the lawyers involved, but each of the residents got a sum as well, typically in the range of one or two thousand dollars. The family of the little girl who had developed asthma got the most in the way of the settlement. It was apparently enough to put a down payment on a new house because the family quickly packed up and moved away.

IN 1989, DESPITE THE BEST OF INTENTIONS on the part of its operators, the Hanahan site was a mess. The sand grains of the aquifer were coated with a sheen of jet fuel, pockets of free product had been trapped under clay lenses, and contaminated ground water was flowing continuously into the neighborhood. Getting the fuel out of the ground was a technical impossibility, and nobody could think of a way of keeping the contaminated water from flowing into the neighborhood. Just when the situation seemed totally hopeless, however, the site operators' luck began to change.

The South Carolina Department of Health and Environmental Control had told the operators, in no uncertain terms, to clean up the site. The person handed this unpromising task was a tall, prematurely graying, and very competent hydrologist named Don Vroblesky. Vroblesky had one thing very much in common with the previous site operators: he had every intention of doing the right thing. But he had something else as well—fifteen years of experience dealing with the twists, turns, blind alleys, and mysteries of ground water contaminant problems. He knew that appearances could be very deceiving in cases like this, and so he and his team of hydrologists and technicians went about the job very carefully.

Any doctor will tell you that the trick to curing a sick patient is to accurately diagnose the illness. And any doctor will tell you that can be very tricky indeed. Vroblesky's team approached the ground water contamination at Hanahan in the same way. Before they did anything, they were going to make sure they *knew* what the problems and pitfalls were. So they began by taking a long, hard look at what had happened over the history of the site.

And what they saw was puzzling. From carefully reviewing the hydrological data, they knew that ground water was flowing into the neighborhood at a rate of about one foot per day. A little arithmetic indicated that the plume of contaminated water should have spread about 5,000 feet away from the site in the fourteen years since the actual spill. But, curiously enough, the contamination had spread less than 500 feet. Why? Clearly, some process was acting to confine the contaminated ground water.

An equally careful review of the scientific literature suggested what might be going on. The plume of contaminated ground water contained mainly soluble BTEX compounds. These compounds, while potentially toxic to humans, were actually pretty good food for the microorganisms that naturally lived in the aquifer. Microbes living in ground water systems don't generally have an overabundance of food. So the sudden appearance of the BTEX actually came as a boon to the local microbial population, which immediately went to work eating the BTEX compounds and transforming them into harmless carbon dioxide. It was the metabolism of these helpful microorganisms, Vroblesky's team discovered, that was confining the BTEX plume.

That certainly was good news to many people in the neighborhood, but it was scant comfort to the people nearest the site, where the plume happened to reach. But Vroblesky's team didn't stop there. The size of the plume, they reasoned, was controlled by two factors: the rate that the BTEX contaminants were being transported by flowing ground water, and the rate that the hungry microorganisms consumed the BTEX. If the rate of BTEX transport into the neighborhood could be lowered, and the rate of microbial metabolism increased, the plume should shrink. And if they could get it to shrink enough, it would cease to be a problem for the unhappy people in the neighborhood.

Pumping and treating contaminated ground water was an accepted technology in 1989. By placing a series of ground water recovery wells around the perimeter of the site, and by pumping an appropriate amount of water, the flux of BTEX-laden water moving into the neighborhood could be decreased substantially.

That was the easy part. The hard part would be to somehow increase the rate microbes were eating the BTEX compounds. One way to do this would be by injecting live BTEX-eating microbes directly into the contaminated plume. This, in fact, was a popular technology back in 1989. But Vroblesky was skeptical. He knew the microbes needed two basic things in order to live: a source of food (carbon and energy), and a source of oxygen. The Hanahan microbes were getting plenty of food in the form of the BTEX compounds.

The problem was, they weren't getting enough oxygen. If they tried to pump in more microbes, those microbes would promptly suffocate and be no help at all. The trick to speeding up microbial metabolism was not to add more microbes, it was to provide those already there with more oxygen.

But that would be difficult. Molecular oxygen (O_2) is only slightly soluble in water, which will hold only about eight parts of oxygen per million parts of water. So it wasn't really feasible to pump in oxygen-laden water. But Vroblesky's team knew something else. Although molecular oxygen is the only kind of oxygen humans can respire, microorganisms are much more flexible. They can also respire the oxygen in dissolved nitrate (NO_3), in dissolved sulfate (SO_4), and in dissolved bicarbonate (HCO_3). Of these possibilities, nitrate was respired nearly as efficiently by the microbes as molecular oxygen, and it was also much more soluble in water. The best way to speed up the overall rate of microbial metabolism in the contaminated aquifer, and thus the rate that BTEX was destroyed, would be to provide the microorganisms with nitrate.

The plan the team came up with was to slow down the delivery of BTEX to the neighborhood by ringing the site with a series of pumping wells, and to speed up the metabolism of BTEX by adding nitrate and oxygen to the ground water through an infiltration gallery. It took a good two years of patient, careful effort to design and build the system. There were regulators to deal with, permits to be obtained, contracts to write, bid, and administer, and a large number of hopelessly incompetent contractors to endure. But Vroblesky persisted, and finally the system was in place.

Early in 1992, the system started up and Vroblesky's team began the long process of monitoring the site in order to assess its effectiveness. This involved sampling about sixty monitoring wells every three months and tracking how contaminant levels changed. For the next two and half years, the team watched with satisfaction as levels of BTEX decreased steadily. By the middle of 1994, concentrations of benzene went from more than 400 parts per billion (ppb) to undetectable levels in some observation wells (figure 23.3A). The impact on xylene concentrations was even more dramatic, with levels falling from more than 2,000 ppb to less than 100 ppb (figure 23.3B). The dual hydraulic containment-bioremediation system was working. The amount of contaminants escaping the site and reaching the neighborhood had been dramatically reduced.

THE ENGINEERS WHO FIRST WORKED on the jet fuel spill can hardly be faulted for their intentions. They sincerely wanted to limit contamination caused by the JP-4 spill, and they went to a good deal of

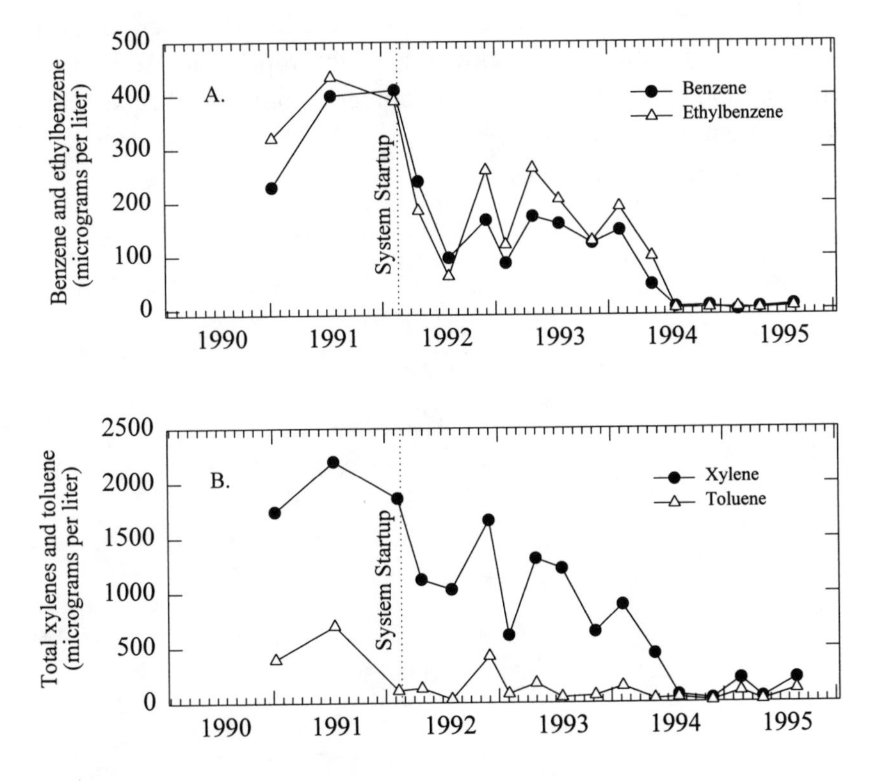

Figure 23.3 Decline of ground water contamination at the Hanahan site when the dual hydraulic containment–bioremediation system began operating.

trouble to accomplish this. What they lacked was solid knowledge about how ground water systems behaved and how they interacted with contaminants. As a result, instead of helping clean up the spill, they managed to make it substantially worse. Don Vroblesky and the folks working with him also had good intentions. But they had something that was just as important—the knowledge, skill, and patience to properly analyze the problem and to work out a feasible solution.

It is tempting sometimes, when faced with huge and seemingly intractable ground water contamination problems, to simply throw up our collective hands in despair. Cleaning up seriously contaminated ground water is such a difficult, laborious, and time-consuming enterprise that it can be very daunting. However, as Don Vroblesky and others have shown, it is not impossible. And while the solution to these complex environmental problems begins with good intentions, there has to be more.

There has to be knowledge.

Chapter Twenty-Four

The Well of Wisdom

CONTAINING AND REMEDIATING GROUND WATER contaminated by jet fuel at the Defense Fuel Supply Center site in Hanahan, South Carolina, consumed a good twenty years of honest effort and an enormous amount of money. But in the end, after numerous costly mistakes, they got it right. That, of course, is the usual progression of things. Most people understand that solving difficult problems requires knowledge that must be paid for step by painful step. This basic fact of life is well preserved in various folktales and mythologies.

One such story concerns Odin, chief of the gods in Norse mythology. Odin carried a heavy burden, for his gods—the Aesir—were in constant battle with the giants of the world. Furthermore, Odin knew that the giants would eventually win and all heaven and earth would be destroyed. It was Odin's sole responsibility to delay this apocalypse—which the Norsemen called Ragnarok—as long as possible.

And so, while the other gods feasted and drank mead, Odin sat and brooded. Even though he was the wisest of the gods, he knew there was much more that he could learn. Accordingly, he rose from the banquet and descended to a secret place that held the Well of Wisdom. To drink water from this well was to gain as much wisdom as was apportioned to gods or men. The well was, however, jealously guarded by Mimir the Wise, who alone among the gods could dispense the sacred water. Odin approached Mimir humbly, as befitted one aspiring to knowledge and wisdom, and begged for a drink of water.

Mimir considered the request, and answered that Odin must pay for the drink by putting out one of his eyes. After due consideration, Odin decided that his eyesight was useless unless he could understand what he was seeing. He consented to Mimir's grisly condition, received his drink of the well water, and took possession of the wisdom that accompanied it. With this new wisdom, Odin conceived a bold new strategy for holding off the giants. He would give knowledge to mortal humans, lifting them above the animals. In return, these mortals would help Odin in his unending battle against the giants.

THE OBSERVATION THAT KNOWLEDGE must always be paid for is not just allegory. Whether you consider TCE contamination of municipal wells in Woburn, Massachusetts, saltwater intrusion into the Floridan aquifer at Hilton Head Island, land subsidence in Phoenix, selenium contamination in the San Joaquin Valley, water level declines in the Ogallala aquifer, nitroaromatic contamination at Weldon Spring, radionuclide contamination in the Snake River Plain aquifer, or jet fuel contamination at a tank farm in South Carolina, the simple fact is that human use of these aquifer systems has created costs. And in some cases these costs have been every bit as grisly as Odin's eye.

But as we have paid these costs, we have also gained knowledge. No longer are we dependent upon forked branches in the hands of water witches to find ground water. No longer need we simply guess at how much water can be safely drawn from a well. No longer need we blindly fear the seeping poisons of chemically contaminated ground water. Now, after centuries of trial and error, after learning the lessons of overpumping, after discovering the folly of using aquifers as waste repositories, and after learning to control the wastes we produce, we have paid our collective eye.

We have earned the wisdom of the Hidden Sea.

Technical Sources

Ackerman, D. J. 1996. *Hydrogeology of the Mississippi River Valley alluvial aquifer, south central United States.* U.S. Geological Survey Professional Paper 1416-D. 56 pp.

Anderson, T. W., G. E. Welder, G. Lesser, and A. Trujillo. 1988. Region 7: Central Alluvial Basins. In W. Back, J. S. Rosenshein, and P. R. Seaber, eds., *Hydrogeology*, pp. 81–86. *The geology of North America*, vol. 0-2. Boulder, CO: Geological Society of America.

Back, W. 1966. *Hydrochemical facies and ground-water flow patterns in the northern part of the Atlantic Coastal Plain.* U.S. Geological Survey Professional Paper 498-A. 42 pp.

Bennett, G. D. 1976. Introduction to ground water hydraulics. In *Techniques of water-resources investigations of the U.S. Geological Survey*, vol. 3, chap. B2. 172 pp.

Bradley, P. M., and F. H. Chapelle. 1995. Factors affecting microbial 2,4,6-trinitrotoluene mineralization in contaminated soil. *Environmental Science and Technology* 29:802–806.

Bradley, P. M., F. H. Chapelle, J. E. Landmeyer, and J. G. Schumacher. 1994. Microbial transformation of nitroaromatics in surface soils and aquifer materials. *Applied and Environmental Microbiology* 60:2170–2175.

Brown, G. F. 1947. *Geology and artesian water of the alluvial plain in northwestern Mississippi.* Mississippi Geological Survey Bulletin 65. 424 pp.

Cederstrom, D. J. 1946. Genesis of groundwaters in the coastal plain of Virginia. *Economic Geology* 41(5): 218–245.

Chapelle, F. H., P. M. Bradley, D. R. Lovely, and D. A. Vroblesky. 1996. Measuring rates of biodegradation in a contaminated aquifer. *Ground Water* 34:691–698.

Chapelle, F. H., and T. M. Kean. 1985. *Hydrogeology, digital solute-transport simulation, and geochemistry of the Lower Cretaceous aquifer system near Baltimore, Maryland.* Maryland Geological Survey Report of Investigations 43. 121 pp.

Cooper, H. H. 1964. A hypothesis concerning the dynamic balance of fresh water and salt water in a coastal aquifer. In *Seawater in coastal aquifers*, pp. C1–C12. U.S. Geological Survey Water-Supply Paper 1613-C.

Darcy, H. 1856. The water supply of Dijon. In R. A. Freeze and W. Back, eds., *Physical hydrogeology*, pp. 14–20. Benchmark Papers in Geology, no. 72. Stroudsburg, PA: Hutchinson Ross.

Deveral, S. J., and R. Fujii. 1988. Processes affecting the distribution of selenium in shallow ground water agricultural areas, western San Joaquin Valley, California. *Water Resources Research* 24:516–524.

Eddy, C. A., B. B. Looney, J. M. Dougherty, T. C. Hazen, and D. S. Kaback. 1991. *Characterization of the geology, geochemistry, hydrology, and microbiology of the in-situ air stripping demonstration site at the Savannah River Site.* Westinghouse Savannah River Company Report WSRC-RD-91-21. 118 pp.

Edmond, J. M., D. L. Von Damm, R. E. McDuff, and C. I. Measures. 1982. Chemistry of hot springs on the East Pacific Rise and their effluent dispersal. *Nature* 297:187–191.

Gutentag, E. D., F. J. Heimes, N. C. Krothe, R. R. Luckey, and J. B. Weeks. 1984. *Geohydrology of the High Plains aquifer in parts of Colorado, Kansas, Nebraska, New Mexico, Oklahoma, South Dakota, Texas, and Wyoming.* U.S. Geological Survey Professional Paper 1400-B. 63 pp.

Heath, R. C. 1984. *Ground-water regions of the United States*. U.S. Geological Survey Water-Supply Paper 2242. 78 pp.

Heath, R. C. 1989. *Basic ground-water hydrology*. U.S. Geological Survey Water-Supply Paper 2220. 84 pp.

Hobba, W. A., Jr., D. W. Fisher, F. J. Pearson, Jr., and J. C. Chemerys. 1979. *Hydrology and geochemistry of thermal springs of the Appalachians*. U.S. Geological Survey Professional Paper 1044-E. 36 pp.

Laney, R. L., R. H. Raymond, and C. C. Winnika. 1978. *Maps showing water-level declines, land subsidence, and earth fissures in south-central Arizona*. U.S. Geological Survey Water-Resources Investigations Report 78-83. 2 sheets.

Lewis, B. D., and F. G. Goldstein. 1982. *Evaluation of a predictive ground-water solute-transport model at the Idaho National Engineering Laboratory, Idaho*. U.S. Geological Survey Water-Resources Investigations Report 82-25. 71 pp.

Logan, R. W., and G. M. Euler. 1989. *Geology and ground-water resources of Allendale, Bamberg, and Barnwell Counties and part of Aiken County, South Carolina*. South Carolina Water Resources Commission Report 155. 113 pp.

Mann, L. J., and L. D. Cecil. 1990. *Tritium in ground water at the Idaho National Engineering Laboratory, Idaho*. U.S. Geological Survey Water-Resources Investigations Report 90-4090. 35 pp.

Meinzer, O. E. 1923. *The occurrence of ground water in the United States, with a discussion of principles*. U.S. Geological survey Water-Supply Paper 489. 321 pp.

Prowell, D. C. 1990. *Geology near a hazardous-waste landfill at the headwaters of Lake Marion, Sumter County, South Carolina* U.S. Geological Survey Open-File Report 90-236. 37 pp.

Randall, A. D., R. M. Francis, M. H. Frimpter, and J. M. Emery. 1988. Region 19: Northeastern Appalachians. In W. Back, J. S. Rosenshein, and P. R. Seaber, eds., *Hydrogeology*, pp. 177–187. *The geology of North America*, vol. O-2. Boulder, CO: Geological Society of America.

Robertson, J. B. 1974. *Digital modeling of radioactive and chemical waste transport in the Snake River Plain aquifer at the National Reactor Testing Station, Idaho.* U.S. Geological Survey Open File Report IDO-22054. 41 pp.

Schumacher, J. G., C. E. Lindley, and F. S. Anderson. 1992. Migration of nitroaromatic compounds in unsaturated soil at the abandoned Weldon Spring Ordnance Works, St. Charles County, Missouri. In *Proceedings from the Sixteenth Annual Army Research and Development Symposium*, CETHA-TS-CR-92062, pp. 173–192.

Seaber, P. R., J. V. Brahana, and E. F. Hollyday. 1988. Region 20, Appalachian plateaus and valley and ridge. In W. Back, J. S. Rosenshein, and P. R. Seaber, eds., *Hydrogeology*, pp. 189–200. *The geology of North America*, vol. O-2. Boulder, CO: Geological Society of America.

Smith, B. S. 1994. *Saltwater movement in the Upper Floridan aquifer beneath Porth Royal Sound, South Carolina.* U.S. Geological Survey Water-Supply Paper 2421. 40 pp.

Speiran, G. K., and W. R. Aucott. 1991. *Effects of sediment depositional environment and ground-water flow on the quality and geochemistry of water in aquifers in sediments of Cretaceous age in the Coastal Plain of South Carolina.* U.S. Geological Survey Open-File Report 91-202. 79 pp.

Theis, C. V. 1935. The relation between the lowering of the piezometric surface and the rate and duration of discharge of a well using ground-water storage. *American Geophysical Union Transactions*, pt. 2, August 1935: 519–524.

U.S. Environmental Protective Agency (EPA). 1995. *Abstracts of remediation case studies*. EPA-542-R-95-001. 129 pp. See especially pp. 58–59.

U.S. Geological Survey. 1984. *Hydrologic events, selected water-quality trends, and ground-water resources*. U.S. Geological Survey Water-Supply Paper 2275. 467 pp.

Vroblesky, D. A. 1992. *Hydrogeology and ground-water quality near a hazardous-waste landfill near Pinewood, South Carolina.* U.S. Geological Survey Water-Resources Investigations Report 91-4104. 87 pp.

Vroblesky, D. A., J. F. Robertson, M. D. Petkewich, F. H. Chapelle, P. M. Bradley, and J. E. Landmeyer. 1996. *Remediation of petroleum-hydrocarbon contaminated ground water in the vicinity of a jet-fuel tank farm, Hanahan, South Carolina.* U.S. Geological Survey Water-Resources Investigations Report 96-4251. 101 pp.

Vroblesky, D. A., J. F. Robertson, and L. C. Rhodes. 1995. Stratigraphic trapping of spilled jet fuel beneath the water table. *Ground Water Monitoring and Remediation* 15:177–183.

Williamson, A. K., D. E. Prudic, and L. A. Swain. 1985. *Ground- water flow and compaction in the regional aquifer system of the Central Valley of California, U.S.A.* U.S. Geological Survey Professional Paper 1401-D. 127 pp.

ABOUT THE AUTHOR

FRANCIS H. CHAPELLE, PH.D., is a research hydrogeologist with the U.S. Geological Survey in Columbia, SC. His work has focused on how microorganisms affect ground-water chemistry in pristine and contaminated aquifers. He is the author of the textbook, *Ground-Water Microbiology and Geochemistry*, and in 1996 his research received a National Award for Environmental Sustainability given by Renew America, Washington, DC.